KU-393-189

A Philosophy of Mathematics

A Philosophy of Mathematics

By Louis O. Kattsoff

Associate Professor of Philosophy
University of North Carolina

Essay Index Reprint Series

BOOKS FOR LIBRARIES PRESS
FREEPORT, NEW YORK

Reprinted 1969 by arrangement

STANDARD BOOK NUMBER:
8369-1086-9

LIBRARY OF CONGRESS CATALOG CARD NUMBER:
73-84314

PRINTED IN THE UNITED STATES OF AMERICA

PREFACE

In glancing back over the development in our understanding of mathematics, we are impressed with the tremendous distance we have come. Contemporary discussions of the nature of mathematics tend to take one of two forms: (a) epistemological, and (b) elemental. Elemental investigation into mathematics had for the most part itself become another branch of mathematics. We can say that the impetus to an investigation of mathematics along elementalistic lines really got under way with Hilbert's metamathematics and Russell's attempt to deduce mathematics from (logic or) logistics. The subsequent unification of these two approaches (never really different) has produced a literature in the field which is growing larger and more difficult of comprehension. The increase in the number of symbolisms, the different techniques employed have caused confusion. Today, *mathematics* is not certain whether or not it should embrace *elementalistics* as a branch of itself, or to leave it to *logistics*—and *logistics* cannot answer this question either. As a consequence, the *Journal of Symbolic Logic* has appeared to offer a medium for publication of results in this field.

The logistic analysis of mathematics has developed into this "elemental mathematics"—a new, and it appears to be, most fundamental branch of mathematics. Elemental mathematics purports to be an analysis of the root concepts such as number, function, or operator involved in mathematics. It appears to out-metamathematicize metamathematics. There is the increasing danger that elemental mathematics is taking itself outside the study of the philosophy of mathematics. But be that as it may, it is casting a great deal of light upon questions in the philosophy of mathematics. Perhaps these questions are meaningless, as logical positivism would have us believe. But even if they are, their very meaninglessness is of interest, for we may still ask the nature, form, and criteria of meaningless questions.

But meanwhile, what of the student who is beginning the study of mathematical philosophy? There are at present no elementary introductions to this field. The student must dive into the heart of things and after struggling some years, comes to one of two conclusions: (a) it's all nonsense, (b) it's beyond him. In either case, he fails to see the depths of the subject and the light it throws on many problems both philosophical and mathematical. Russell's *Introduction to Mathematical Philosophy* is a summary of results (p. v) rather than an introduction. Black's *Nature of Mathematics* is an excellent discussion for those already in the field. The many smaller pamphlets such as Heyting, Weyl, Dubislav, Tarski, and others have issued, are too brief and difficult or else a treatment of special problems such as *Der Zahlbegriff seit Gauss*, *Das Unendliche in der Mathematik*. As for the more epistemological questions, these are scattered through the works already mentioned and through many others too numerous to mention.

This book is an attempt to give the beginning student an introduction to the many problems raised by the "queen of the sciences." It attempts also to gather together material scattered throughout the literature and arrange this material to show the development of the ideas in the mind of its creator. Its final aim is to present some of the implications of these problems and proposed solutions for philosophy in general. Naturally, it will not cover all problems and cannot even attempt to exhaust the problems it does discuss. Most of the problems treated will be related to what is called the more elementary branches of mathematics. Even so, students should be fairly well acquainted with mathematics through the functions of a real variable, at least, before attempting to speculate upon these problems. In any case, a decided maturity in mathematical thinking is presupposed.

This book is based upon courses in philosophy of mathematics given at the University of North Carolina to students whose knowledge of mathematics was that of an undergraduate senior and whose philosophy was of equal order. These were students whose major interest lay in mathematics or philosophy.

LOUIS O. KATTSOFF

November, 1947

ACKNOWLEDGEMENTS

The author wishes to acknowledge his indebtedness to the many people who have aided in the writing and publication of this volume. He especially wishes to acknowledge a student's debt to his former teacher, Henry Bradford Smith, under whose guidance interest in these studies was first developed. Appreciation is also due those who have read and criticized various parts of the manuscript, especially Professor H. S. Leonard of Duke University and Professor W. Hurewicz. Thanks are due to my wife for her invaluable aid in typing and retyping the material. And to Professor Edward S. Allen of Iowa State College I am indebted for his painstaking and careful reading of the entire manuscript. His many suggestions have made it decidedly better.

L. O. K.

TABLE OF CONTENTS

CHAPTER ONE

PHILOSOPHY AND MATHEMATICS

Mathematics, as practically all the sciences, has made tremendous advances since the turn of the century. Its triumphant advance appears to be unabated. The increase in our mathematical knowledge has made it practically impossible for any one man to be the technical master of all branches of this science. New fields of mathematics have been opened; some of these will be indicated later in our attempt to indicate how the extension of the mathematical system can be interpreted. With the increase in the technical side of mathematics, there has come a steady increase of interest in the relations of this science to the rest of our world; in the significance of the fundamental concepts from which this edifice is constructed; in the study of the methods of proof which give to this science its apparent certainty; in the nature of this certainty; in the very nature of mathematics itself.

There are two avenues of approach open to one who desires to study mathematics. One is to enter some special field, as theory of functions, or topology, and to advance the details within that field. This is pure mathematics. The second is to trace the origins of mathematics to their roots and to study those. This gives rise to the types of questions mentioned above. This is the philosophy of mathematics. But obviously each is necessary to the other. The study of pure mathematics leads to a better knowledge of the roots of the science. The study of the structure of mathematics leads to a greater facility in pure mathematics. Furthermore, the complete structure helps the study of other fields of knowledge. It is no wonder then that Leibniz, who was both a philosopher and a mathematician, said, "Mathematicians have as much need of being philosophers as philosophers have of being mathematicians."[1]

The mutual benefits to be derived by each from the other have lessened to a great degree the acrimony with which mathematicians have attacked philosophers, and conversely. Most of this ill feeling has been due to a great degree to the attempt on the part of each group to keep themselves free from the other.

The philosopher interested in mathematics does not concern himself to any degree with the discovery of new facts, except insofar as he is dealing with elemental mathematics. But he is interested in these

[1] G. Leibniz, "Letter to Malebranche," *1699 Philosophische Werke* (Gerhardt ed.), Vol. I., p. 356.

new facts since they do constitute the elements out of which the mathematical structure is erected and to which all theory must be subjected. The discovery of a general solution for quintic equations is not a matter for the mathematician or philosopher interested in philosophy of mathematics—although the existence or nonexistence of such a solution, or the methods of deriving these solutions, may be.

The problems which the philosopher interested in mathematics considers are divided into two main sections: (1) metamathematical, and (2) epistemological.[2] Under metamathematical (a term derived from Hilbert,[3] which will become clearer later) we have the following examples: the problem of the consistency of mathematics; the problem of the universality and applicability of mathematics; the problem of the completeness of mathematics; etc. These problems were very carefully investigated by David Hilbert. These problems are concerned with properties of postulate sets. (An illustration of a postulate set is the set of axioms we all learned in our elementary, Euclidean geometry.) However, since it is now thought that all mathematics should ultimately be reduced to postulational form, these metamathematical problems really involve all mathematics. Metamathematical investigation is concerned with problems relating to the skeletal structure of mathematics as a structure. It concerns itself with those relations between the symbols and symbolic combinations in mathematics, which constitute the "grammar" of the mathematical "language." Under epistemological (i.e., referring to the "theory of how we know"), we find such problems as the method of establishing mathematics, the method of constructing definitions in mathematics, the general nature of mathematics, the so-called "object of mathematics," the nature of existence in mathematics, the part infinity plays in mathematics, the relation of mathematics to other sciences.

The problems of the first type have given rise to what I have termed "Elemental Mathematics" and have tended more and more to become mathematics rather than the philosophy of mathematics. (Cf. the discussion of Church's work, outlined in Chapter 10.)

Another classification of the problems involved in a philosophy of mathematics is given as:

1. The content of mathematics and its evolution.
2. The central principles of mathematics.
3. The source of reality of mathematics.
4. The methods of mathematics.
5. The regions of the validity of mathematics.[4]

---

[2] W. Dubislav, *Die Philosophie der Mathematik in der Gegenwart* (Berlin, 1932), p. 1.

[3] Cf. David Hilbert, "Die Logischen Grundlagen der Mathematik," *Math. Annalen* Vol. 88(1923), pp. 152–53.

[4] J. B. Shaw, *Lectures on Philosophy of Mathematics* (Chicago, 1918), p. 6.

This classification is, however, inadequate, since it restricts itself for the most part to epistemological problems.

One of the very important problems which must be mentioned is that of the nature of symbols which has come to be called the semiosis problem.[5] This problem has two aspects which may be denoted by the names given above: (1) metamathematical, and (2) epistemological. Under its metamathematical aspect we investigate the types of symbols, their functions and formal properties, as they occur in mathematics. Under its epistemological aspect we investigate the general nature of symbols, the use of symbols, the applicability of mathematical symbols to other fields. As a matter of fact, many have insisted that the fertility of mathematics is due to its symbols, and many have dreamt of extending the use of these symbols or of discovering similar symbols for other fields, and hence of reducing other fields to mathematical form. To some extent this dream of Descartes' to establish a "universal science of order and measurement" is being realized. But the question still remains whether such a procedure is capable of being imposed upon all fields of knowledge.

Recent developments in the theory of signs have distinguished three types of problems relating to signs: (a) syntactics—the relations of symbols to each other, (b) semantics—the relation of signs to the objects they designate, (c) pragmatics—the relation of signs to the subject for whom they are signs. Syntactics is a metamathematical sphere. Semantics and pragmatics are in the sphere of epistemology. A valid theory of mathematics cannot separate the two aspects of syntactics and semantics.

Perhaps the best distinction between the types of problems may be defined as follows: The metamathematical problem deals with symbols, which it sorts systematically and uses as subject matter, treating their combinations and properties without considering their meaning. These are syntactical problems. The epistemological problem is "nonformal, concentrates upon the subject matter of which the system treats, and advances by intuitive insight into a recognition of the nature of phenomena."[6] It treats the meaning of the symbols. These are semantic and pragmatic problems.

It is evident that symbols have two types of properties: (1) properties of that which is used as a symbol, (2) properties which they have as a consequence of that which they symbolize. For example, 5 as a mere symbol represents $1 + 1 + 1 + 1 + 1$; as a number it has the properties numbers have. Obviously there are certain types of symbols which

[5] Cf. the discussion of Mannoury (and the literature there mentioned which occurs later). Also the pamphlets of Morris and Carnap in *Int. Encyclopedia of Unified Science, Vol. I.*

[6] M. Black, *Nature of Mathematics* (London, 1933), pp. 141–42.

can best be used to represent definite concepts. This is a problem in the philosophy of mathematics, for it is by means of these symbols that empirical generalization is made possible. It is these symbols which make possible the systematic body of knowledge we call mathematics. And it is the task of critical philosophy to clarify knowledge by criticizing such systems. "Mathematics is a most admirable field for the exercise of applied philosophy."[7]

All of the problems of the philosophy of mathematics so far mentioned, and they can all be subsumed under the two divisions of metamathematical and epistemological, fall naturally into the two divisions of philosophy—critical and speculative, (dealing, respectively, with the analysis and the synthesis of concepts).[8] Mathematics as a science *uses* concepts: number, function, variable, point, axiom, set, group, relation, existence, etc. It does not analyze them. The function of critical philosophy is the analysis of these concepts. This does not mean that the mathematician does not *use* these concepts fairly consistently; in many cases he does. But we desire to know *about* these concepts, even if we discover ultimately that the only meaning they have is the way they are used, i. e., that the only definition we can give of them is a "*Gebrauchsdefinition.*"

These concepts are usually expressed in propositions which are assumed as the foundation of the science. It is the task of critical philosophy to analyze the types of propositions, their relation to each other (consistency, independence, categoricity, etc.). Only comparatively recently have mathematicians come to see the tremendous importance of these latter problems. The analyses of concepts and of propositions go hand in hand. The properties of the concepts are learned by studying their occurrences in propositions. The forms of the propositions are determined by the type of concept which occurs in them. The analysis of concepts and propositions has long been the task of logic. It can therefore easily be seen that mathematics and logic have some relation. The problems of a critical philosophy of mathematics are the metamathematical problems of mathematics.

No science which is still progressing is more than partially self-consistent. "For scientific research is characterized by the choice between mutually inconsistent theories. . . ."[9] This indicates two things: (a) at any given time we may have a self-consistent subsystem of a larger system; (b) we may have two systems, each of which is self-consistent, but which are incompatible with each other. Case "a" necessitates the development of the larger system, and that is a matter of

---

[7] Black, *op. cit.*, p. 1.

[8] C. D. Broad, Critical and Speculative Philosophy in *Contemporary British Philosophy* (First Series, London, 1924), pp. 82 ff.

[9] Black, *op. cit.*, p. 2.

technical mathematics. Illustrations of such procedures occur in Euclid's development of geometry, and, in contemporary mathematics, in R. L. Moore's "Foundations of Point Set Theory."[10] Each chapter in Moore's text develops theorems from a number of axioms and contains at least one axiom not in the preceding chapter. Case "b" is illustrated by Euclidean and one of the non-Euclidean geometries. In this case we have the additional problems of determining the source and nature of the incompatibility, the significance of such systems for the question of the applicability of mathematics, the implications for our body of knowledge, etc. These are all tasks of speculative philosophy. They are also among the problems of epistemology mentioned above. The Pythagoreans of Ancient Greece, Descartes of the seventeenth century, and Whitehead of the twentieth century are illustrations of the part mathematics plays in speculative philosophy and of the speculative philosophy of mathematics.

Critical philosophy of mathematics may be said to have as its task the attempt to determine whether or not mathematics is *logical*; speculative philosophy whether or not mathematics is *rational*.[11] This, of course, introduces a distinction between what is logical and what is rational. Thus, there may be many systems of concepts all equally logical (i. e., consistent and implicative) but mutually incompatible, of which only one would be rational, i. e., applicable to reality. In order that we have knowledge our system must be both logical and rational. Thus, the theory of differential equations may be logical but it becomes rational also, and then gives us knowledge, only when it is applied to physics; i. e., mathematical physics (within limits) is a branch of knowledge both logical and rational. It was this that Descartes meant by the universal science of order and measurement.

Neither critical nor speculative philosophy attempts to deny or destroy mathematics; rather, the purpose of the philosopher is to understand and to establish firmly this "queen of the sciences." The investigation of the foundations (logistic, intuitionist, and formalist) has lead to the so-called "Erschütterung" of mathematics (cf. the work of Brouwer later). It has been maintained that the entire structure of mathematics was weak or at least built upon weak foundations. Too often thinkers attempt to justify their position by a negative attitude toward what went before! The investigation of the foundations of mathematics is of interest in itself even if it does not destroy these foundations. The structure of mathematics has been singularly un-

---

[10] American Mathematical Society Colloquium Publications, (New York, 1932), Vol. XIII.

[11] This distinction is attributed by Couturat to Cournot. Cf. L. Couturat, *De l'infini mathématique*, (Paris: Alcan, 1896), p. x. Contemporary speculation, based on Wittgenstein, tends to overlook the possibility that a system may be both logical and rational.

affected by the controversies. The foundations are being inserted without much damage. This is to be expected, even though it may alter our insights into the nature of mathematics. Even the relativity theory, which has changed the entire approach to physical problems has not altered and cannot alter the facts gathered by experiment. But the Grundlagenforschungen (researches on the foundations) have caused us to change our concept of rigor in proof,[12] have given us a better insight into the nature of the infinite in mathematics, have demonstrated the self-contradiction of certain concepts which were being used in the newer branches of mathematics, have developed certain paradoxes as a result of these contradictory concepts, have discovered the Zermelo axiom (see later), etc. All of these facts are sufficient evidence of the importance of these investigations without imagining that these investigations have caused the edifice of mathematics to totter.

We see that philosophy has two main divisions of problems: the critical and the speculative; and the two divisions of problems in the field of the philosophy of mathematics—the metamathematical and the epistemological—are subsets of these major divisions. This means that the metamathematical problems are critical problems relating to mathematics and the epistemological ones are speculative problems relating to mathematics.

As a result of our discussion, it is clear that the philosophy of mathematics is of interest (1) to mathematicians desiring to see the larger significance of their science and to get an insight into its structure, and (2) to philosophers interested in getting an insight into the fundamental concepts of mathematics, into the application of logic to a specific science, and into the light which this great field of human knowledge can shed upon epistemological and metaphysical problems. The development of interest in the philosophy of mathematics is but one bit of evidence of the growing "rapprochement" between the various sciences and between science and philosophy.[13] This is an excellent tendency since it leads to a more definite "Weltanschauung" which is so badly needed today.

The point of view taken in this book is definitely determined by certain metaphysical doctrines. Our attitude towards the nature of mathematics and some of its implications for the nature of reality, we shall indicate at the end of our discussion, after we have presented the facts upon which it is based. The approach to these problems is not

---

[12] Cf. E. T. Bell, "Place of Rigor in Mathematics," *American Mathematical Monthly*, Vol. 41, (1934), p. 599.

[13] Russell points out that the distinction between philosophy and mathematics is one of point of view. This follows from his view as to the nature of mathematics. (*See* Chap. 2, B. Russell, *Principles of Mathematics*, p. 129).

to be based on pseudo-pedagogical doctrines that insist that we must go from the simple (to the reader) to the complex (to the reader), but rather on the idea of beginning with the roots and developing the structure. Since it is assumed that readers will have some mathematical maturity, we shall first examine definitions of mathematics. It is a fact, however, that these will become clearer and more meaningful as the work progresses.

Definitions are always tentative expressions of one's insights up to that point. Since definitions delimit the object of a given field, some discussion of the nature of the object of mathematics is in order at this point. Here again, the significance of much of the discussion and understanding of the nature of the mathematical object will become clearer at the end of the book. But we need some ideas with which to begin. In any case, the most prevalent idea of the object of mathematics is that mathematics deals with number. To most people unacquainted with much mathematics this is still the case. Number is, perhaps, the simplest mathematical concept found in the system of mathematics as ordinarily conceived. Many have felt that from it all the rest of mathematics can be constructed.

A discussion of definitions of number and the extension of the number system, accordingly, is in order. We shall attempt to describe the lines of development that occurred separately and were joined later. The extension of the number system leads to a discussion of the paradoxes related to the "transfinite" numbers, and as a consequence to the attempts to reconstruct the foundations of mathematics to remove these paradoxes. Foundational investigations lead to the discussion of consistency and independence proofs, and the development of the idea of logical syntax. This line results in the discoveries of Gödel, which are of extreme importance. With these topics we have before us the investigations of the complete structure of mathematics and, therefore, of postulational systems and their interpretations. Since interpretations of postulate sets are really the applications of mathematics, we are at the problem of the relation of mathematics and reality and the metaphysics of mathematics.

If we appear to spend too much time with metamathematical problems, it is because of their importance in understanding the nature of mathematics and their implications for epistemological problems.

CHAPTER TWO

DEFINITIONS OF MATHEMATICS

The discussion of the object of mathematics and the attempt to define mathematics come both at the beginning and at the end of a study of mathematics. Obviously, critical philosophy is as necessary to the discovery of a definition of mathematics as is speculative philosophy. It is significant that with the increase in the critical investigation of mathematics, the definitions of this science underwent a change. As a matter of fact, as we shall see during the course of our investigation, definitions of mathematics vary in relation to the type of investigation of the definer. Each definition indicates that aspect of mathematics which the investigator favors. No *final* definition of mathematics has yet been given and perhaps no final definition will ever be given so long as the science remains as fruitful as it has been and still is. But a consideration of the various definitions which have been given will give us some insight into some of the characteristics of the science.[1] The knowledge of these properties of mathematics leads us to a new consideration of the science—which in turn leads us to a reconsideration of the definition given. This would be evident if we were to undertake a study of the historical development of definitions of mathematics parallel to the development and criticism of mathematics. But all of these definitions are of two main types: (1) definitions resulting from a consideration of the critical philosophy of mathematics, i.e., what might be termed metamathematical definitions; (2) definitions resulting from a consideration of the speculative philosophy of mathematics, i. e., what might be termed epistemological definitions. The present tendency is to give critical definitions, definitions based upon method, or subject matter, rather than epistemological definitions. This, I believe, is due to the fact that the increase in mathematical knowledge, foundation theory, and elemental mathematics has outstripped those philosophers who are unacquainted with these fields.

The multiplicity of definitions makes some selection necessary. We have selected what appear to be the most representative of the great numbers of definitions given in the literature. These seem to fall into

[1] M. Bôcher, "Fundamental Conceptions and Methods of Mathematics," *Bull. Amer. Math. Soc.*, Vol. 11, (1904), pp. 115 ff.

three types, two of which are definitely epistemological and the third is definitely metamathematical. We shall consider each group separately.

GROUP 1

(1). Mathematics is the science of magnitudes.

(2). Mathematics is the science of numbers.

(3). Mathematics is the science which has for its object the indirect measurement of magnitudes.

These definitions clearly hark back to the days before mathematics began to develop the various divisions it now has. Leibniz, Gauss, and Euler are among the names of those who held these definitions. And it is to Comte[2] that the third definition can be attributed. Critical remarks concerning these definitions are fairly obvious and come to mind easily. That mathematics deals with more than magnitudes and numbers any tyro in the field knows. Even as early as 1810 this was called to the attention of philosophers by Bolzano[3], to whom we owe our present definition of infinite classes.

But Bolzano felt that the concept of magnitude was central to mathematics and he tried to give it a more precise analysis than had either Gauss or Euler. Gauss had merely divided magnitudes into two kinds: (1) extensive magnitudes, i. e., those having parts (as, e. g., geometric magnitudes), and (2) intensive magnitudes, i. e., those that do not have parts (as, e.g., tones, velocity, etc.). He had then asserted that mathematics considers the relations between magnitudes which they have in virtue of being magnitudes—these give rise to arithmetic—as well as the relations between magnitudes which they have as a consequence of their position or place—these give rise to geometry.[4] But such an explanation really offers very little so far as the properties of magnitudes are concerned. It leaves entirely to one's imagination the meaning of the phrases "relations . . . which they have in virtue of being magnitudes," and also what one could mean by the position or place of a magnitude. Even Euler's comment that a magnitude is something which is capable of increase or diminution or to which something can be added or subtracted, although a step in the direction of greater precision, is vague and unclear. Furthermore, Euler's remark presupposes that we already have a magnitude which increases or diminishes, the given magnitude; that is, as Bolzano had pointed out, magnitudes are really numbers.

Bolzano attempted a definition of magnitude that, in point of view, is

[2] A. Comte, *The Positive Philosophy*, trans. abridged by H. Martineau, (New York, 1854), p. 38.

[3] H. Bergmann, *Bolzanos Beiträge zur philosophischen Grundlegung der Mathematik*, Halle (1909) p. 12.

[4] C. F. Gauss, (1777–1855) *Werke*, (Berlin ed., 1929), Vol. 12, pp. 57–61.

extremely modern. His attempt was to set up a constructive definition; i. e., a definition such that the entities were to be considered magnitudes, if we could build up from one to the other. Consider, for example, 3 and 7. Since by the addition of 4 to 3 you would get 7, 3 and 7 are magnitudes. Similarly since by addition of 1 to 3 we get 4, 3 and 4 are also magnitudes. By addition of 2 to 1 we get 3, so 1 and 3 are magnitudes. And by addition of 1 to 1 we get 2, then 1 and 2 are also magnitudes. In general terms, all "objects any two (say M and N) of which are either equal to each other or such that one of them can be represented as a sum, one term of which is the other, i. e., $M = N$ or $M = N + n$ or $N = M + m$" are called magnitudes (m and n must also satisfy this definition.)[5]

If we permit ourselves a certain amount of freedom in the interpretation of the = and the +, this definition is applicable not merely to numbers, but to groups, invariants, spaces, lines, etc. It would, therefore, appear to fit most objects with which the science of mathematics is concerned.[6] As a matter of fact, if such interpretation is permitted, the definition becomes too broad, i. e., it will include objects that do not belong to the realm of mathematics. We construct an example of such a case.

Consider the class consisting of the following names:

$$C = \{\text{Al, Ali, Alif, Frail, Flair}\}$$

Let M and N be any two names in C. We define $M = N$ to mean that M and N have the same letters. $M = N + n$ means "the name M consists of all the letters of N and in addition some of the letters of another name n." $N = M + m$ is similarly defined. Then for any two names in C, one of the relations $M = N$, $M = N + n$, $N = M + m$ holds. Any name, as a matter of fact, except the first and the last may be so written by taking the name preceding it as N and the name following it as n. The last name is equal to the preceding one; while every name can be written using the first name as N. The elements of C would then need to be magnitudes!

Other similar illustrations are possible. For example; if $C = \{a, ab, abc, abcd, abcde, abecd\}$. The elements need only be formed in a manner similar to numbers, i.e., each element should contain the preceding one and in addition one other. If the sequence is finite then the last element should be "equal" to its predecessor.

The point is that magnitudes do satisfy the given property that one of the relations $M = N$, $M = N + n$, $N = M + m$ be fulfilled, but so do other things besides magnitudes.

---

[5] H. Bergmann, *op. cit.*, p. 57 and also, B. Bolzano, *Paradoxien des Unendlichen* (1851) Leipzig 1920, p. 4.

[6] For a more recent development of number from magnitude, cf. Stoltz, D. und Gmeiner, J. A., *Theoretische Arithmetik*, Leipzig und Berlin, 1911, Vol. I, pp. 1–16.

If restrictions are placed upon the definition so that it applies to numbers only, then definite problems arise. (1) What are these restrictions? (2) Is this definition foundational in character? (3) Can we derive all branches of mathematics from this definition? The answers to these questions will anticipate future discussions and are given only in brief. In the first place, this is actually not a definition but a statement of a property which numbers have. But if it were a definition, it would not be generally valid if the M and N do not belong to the same type; e. g. $\sqrt{-2}$ and 5 are not such that either $\sqrt{-2} = 5$ or $\sqrt{-2} = 5 + n$ or $5 = \sqrt{-2} + m$, unless m and n may be complex numbers. If we consider the entire number system, then this property is that which indicates that the system is "closed." This indicates that the given definition is not foundational in character (we shall see this more clearly later). The third question will also be answered negatively since additional axioms are necessary. This, too, will be evident later.

The fundamental character of number for a great deal of mathematics is beyond dispute, but the definition of number and its roots are not given in these statements.

The centrality of the concept of magnitude did not prevent Bolzano from attempting a totally different kind of definition, which is basically epistemological.

GROUP 2

4. "Mathematics is a science which deals with the universal laws (forms) according to which things must behave as existents.[7]

5. "Mathematics deals with "the investigation of concepts which express the relations of any objects to each other."[8]

For the purpose of formal derivation of the equations of mathematics, such definitions as these are inadequate. They attempt to state the field of application of the formal structure; that is to state the internal relations of mathematics to reality. Bolzano's definition (4) involves an element that was later taken to be a chief characteristic of mathematics—namely, the element of possibility. Mathematics treats not of objects, but of the way in which objects must behave if they are to become phenomena. This makes of mathematics actually a hypothetical science which has nothing to do with existent objects. "Mathematics," says Peirce, "is engaged solely in tracing out the consequences of hypotheses."[9]

---

[7] Bergmann, *op. cit.* p. 15.

[8] E. S. Papperitz, "Über das System der rein mathematischen Wissenschaften," *Jahresb. des Deuts, Math. Verein.* Berlin, 1890–91, Vol. I, p. 36.

[9] C. S. Peirce, *Collected Papers*, Cambridge, 1933, Vol. I, pp. 23, 78, 113, also Vol. III, pp. 346–52.

Both these definitions have their origin in Kantian philosophy. For Kant, mathematics was precisely the science of the forms of thought and this meant that mathematics concerned itself with the forms in which objects appeared to the mind. Kant's approach was through the question of discovering what was necessary for experience to be possible at all. Another approach to the same point of view comes through the recognition of the fact that mathematics seems to be applicable to almost any set of objects, and what these objects all have in common are their laws of appearance, i. e., their regularities as existent objects.

These definitions attempt to express the idea that mathematics is basically a science which describes facts inherent in the structure of reality. These seek to impart to mathematics a necessity which arises from this source. They desire to bridge the gap between formal technical mathematics and the cosmos in which mathematics exists and from which it apparently arises. Mathematics for Bolzano and Papperitz, the authors of these definitions, gives us universal laws. The statements of mathematical propositions are actually statements about reality as perceived; or rather about the properties things must have to be perceivable.

It is easily seen that if one wished to identify the laws of logic with these "universal laws according to which things behave"—and this has been done in the history of logic—these definitions are not far from the identification of logic and mathematics. The actual identification, or attempted identification, was made through a different approach, as we shall see later.

The third group of definitions approach the problem of a definition of mathematics from its more technical aspect. The attempt is made to characterize mathematics in terms of its formal structure apart from either its subject matter or its relations to other phases of reality. These definitions were given mostly by scholars who were pure mathematicians or who were interested in the formal side of logic.

GROUP 3

6. "A mathematical system is any set of strings of recognizable marks in which some of the strings are taken initially and the remainder derived from these by operations performed according to rules which are independent of any meaning assigned to the marks."[10]

7. "If we have a certain class of objects and a certain class of relations and if the only questions which we investigate are whether ordered groups of these objects do or do not satisfy the relations, the results of the investigations are called mathematics."[11]

---

[10] C. I. Lewis, *A Survey of Symbolic Logic*, Univ. of Calif. Press, 1918, p. 355.
[11] Bôcher, *op. cit.*, p. 127.

8. "Mathematics is a pure system (Lehrbegriff)in the sense of a formal theory. It is a science (*W*) which has the following properties:

a) The propositions in *W* contain in addition to purely logical concepts, only variables.
b) No proposition in *W* is a proposition of pure logic.
c) There exists a finite sub-set *A* of *W* (called the initial propositions of *W*) such that the propositions of *A* are compatible with and independent of one another, and from these every other proposition in *W* can be deduced by purely logical deductions. Every proposition, finally, which can be deduced from *A* is a proposition in *W* unless it belongs to pure logic."[12]

9. "Mathematics is the science which draws necessary conclusions."[13]

10. "Mathematics in its widest signification is the development of all types of formal, necessary, deductive reasoning."[14]

11. "Pure mathematics is the class of all propositions of the form '*p* implies *q*' where *p* and *q* are propositions containing one or more variables, the same in the two propositions, and neither *p* nor *q* contains any constants except logical constants. And logical constants are all notions definable in terms of the following: Implication, the relation of a term to a class of which it is a member, the notion of *such that*, the notion of *relation*, and such further notions as may be involved in the general notion of propositions of the above form."[15]

It will be noticed that the general tenor of these definitions is the same. There is not much difference between Korselt's definition and Russell's and certainly Peirce's and Whitehead's definitions are but less precise forms of Russell's.

This kind of definition is today the most popular one. It is the culmination of a number of lines of research. The work of Frege and Boole in attempting to give logic a mathematical form and Frege's attempt to derive mathematics from logic were one set of very important developments. Meanwhile, also, there was increasing arithmetization of mathematics, stimulated by Weierstrass, Dedekind, and others, which tended to reduce irrationals and other concepts to combinations definable in terms of the series 1, 2, 3, . . . (as will be shown later). And added to these were the attempts of men like Hilbert to set up a final postulational system for geometry which led to the attempt to

---

[12] A. Korselt, "Was ist Mathematik?" *Archiv der Math. u. Phys.* 1913, Vol. 21, p. 371.

[13] B. Peirce, (1870) cited by Bôcher, *op. cit.*, p. 117.

[14] A. N. Whitehead, *Universal Algebra*, Cambridge, 1898, p. vi.

[15] B. Russell, *Principles of Mathematics*, Cambridge, 1903, Vol. I., p. 3. But in the introduction to the 2nd edition, W. W. Norton, 1938, p. ix, Russell insists that this definition is too wide and that propositions that belong to mathematics must also be tautological.

throw other branches of mathematics into similar postulational form. Russell's definition derives directly from the work of Frege, whom he rediscovered. Lewis' and Korselt's definitions show the influences of the increasing recognition that a postulational system is a general form and that geometry is an application of such a form to points, lines, etc. As Bôcher[16] (among others) had pointed out, the intuitive element in geometry can easily be removed by a change in language. If for points we use the expression *A*-objects, and for lines, *B*-objects, then the expression "two points determine a line" becomes "two *A*-objects determine a *B*-object." For any two given *A*-objects, there exists one and only one *B*-object to which they both bear the relation *R*." It is possible to give other interpretations to the *A*-object, *B*-object, and *R*, so that we have actually generalized the geometric system and are not talking, in mathematics, about particular instances or particular entities. "The point of mathematics is that in it we have always got rid of the particular instance and even of any particular sorts of entities."[17]

Lewis is especially emphatic about the independence of the meaning from these marks, and views as the important aspect of mathematics the *type of order* present in it. In this he differs from Russell for whom the marks have definite meaning in terms either of logical classes or of logical propositional functions. Lewis even goes so far as to insist that the reason for using *marks* rather than sounds or odors is merely one of convenience.

All of these definitions in Group 3 show the result of the recognition of the close relationship between logic and mathematics. This had apparently been recognized clearly even before Frege had begun the attempt to derive mathematics from logic, since Peirce's definition given about 1870 implicitly makes the identification, and Whitehead in 1898 had explicitly defined mathematics in terms of deductive reasoning. We shall see later some of the difficulties involved in this identification.

The prime difficulty with many of these definitions seems to be that they are too wide. They need restrictions to exclude some fields and to identify mathematics uniquely. For example:

That definition 6 is too wide can be shown thus: Consider a deck of fifty-two ordinary playing cards. These constitute our recognizable marks. We consider that the game of "poker" is to be played. The first deal made of five cards to each player will constitute the set of strings taken initially. The rules of the game are also among these initially taken sets of strings. All remaining sets of strings (i. e., subsequent "hands") are *derived* from the initial set by the rules of the game. ("Casino" is perhaps a better game to use as an illustration.) Actually,

[16] Bôcher, *op. cit.*, pp. 121–22.
[17] A. N. Whitehead, *Science and the Modern World*. Macmillan Co. 1928, p. 31.

the game of chess has been used as an analogy with which to compare mathematics—but no one, to my knowledge, has claimed that chess is mathematics.

The basic distinction would be between the application of mathematical techniques to a specific subject matter and a branch of mathematics. To assert that wherever we can apply mathematical methods we have a branch of mathematics would be to make extremely vague our notions of subject matters. For example, to apply the calculus to problems of dynamics and to problems of economics is not to identify dynamics with economics. Certainly the development of a mathematical theory of games which can be interpreted as an economic theory[18] would not reduce the problems of economics to those of mathematics, nor would it equate (except in analagous fashion) buyers and sellers with mathematical concepts.

Definition 7 is also too wide. Given the class of objects "adult human beings in the town of Chapel Hill," and the class of relations "is related to, is married to, is mistress of, is enemy of." Suppose we investigate the questions whether the individuals obtained by replacing specific persons for the general term in the class defined by (male, female) do or do not have one or more of these relations between them. According to our definition, results of these investigations would be mathematics.

Mathematics is a many-sided science. For many centuries it was the glory of rationalist philosophers and to this day it is an outstanding achievement of human thought. It cannot be uniquely defined by stating its structure or its field of application alone. If it were true that the formal structure of an implement made that implement applicable to one and only one entity, then to state either would be sufficient. But the mathematical structure is applicable to many entities, and, there is reason to believe, to all types of entities. Hence that to which mathematics is applicable would appear to be common to many, if not to all, types of entities. Since any increase in the number of entities makes for more general common properties, it appears that the subject matter of mathematics must be general in nature and perhaps the most general of all concepts or entities.[19] However one finally decides to define mathematics, the two aspects of formal structure and general subject matter must be included.

---

[18] Such a theory has been partially developed by John von Neumann and Oskar Morgenstern, *Theory of Games and Economic Behavior*, Princeton, 1944.

[19] It is interesting to compare these ideas with the viewpoints of Kepler and Galileo. Cf. E. A. Burtt, *Metaphysical Foundations of Modern Physical Science*, New York, 1932, the chapters on Kepler and Galileo.

CHAPTER THREE

THE NATURE OF THE OBJECTS OF MATHEMATICS

In any discussion of the objects of mathematics, we must first carefully separate, in our minds, mathematics and the mathematical treatment of certain entities. This means that we must distinguish between two semantical objects. If we view the object which mathematics studies to be merely signs and their syntax, then this is one semantical object. In studying what is commonly known as pure mathematics, the semantical object of the study is the sign itself. On the other hand, if we consider the application of mathematics to a specific field, then the semantical object of mathematics is that to which the sign refers, e. g., to motion, to time, etc.

When we come to discuss the abstract character of mathematics, it will become evident that mathematics is a mold—a form—a system—which is applicable to many entities. It is almost possible to say that mathematics is a language and that its elements (groups of symbols) are the "words" of the language. It would then appear to be possible to analyze mathematics linguistically.[1] Such a point of view explains the different definitions of mathematics and the reason it has taken us so long to see the nature of mathematics. As in our ordinary language a word may stand for a group of entities but not for *all* things, so in mathematics each symbol represents a mathematical entity or thing. For example, $\frac{dx}{dy}$ represents one meaning, while $f(x)$ represents another, and 5 represents still another. Likewise, words which have a meaning in our ordinary language have a different meaning in mathematics. Thus *space* has a psychological meaning and also a mathematical meaning.

It is possible that the mathematical meanings of words are abstractions from the meanings the words have in other fields. Certainly, if mathematics is a form in which other branches of knowledge can be thrown, the symbols or objects which occur in it must represent skele-

[1] Cf. A. F. Bentley, *A Linguistic Analysis of Mathematics*, Indiana, 1932. Also Black, *op. cit.*, pp. 129–34. Mathematics had been considered the "syntax" of symbolic languages. Also for criticism, Croce, *Logic*, London, 1917, Chap. 6.

tons of things. It is in this sense that we can speak of the symbols of mathematics as *variables*. But each variable represents an object which is analagous to the steel framework of a building. It is possible to put around the framework various forms, but the range of these forms is determined by the framework. Mathematics, then, has these variable-objects (or framework-objects) as its subject matter, while the mathematical treatment of entities limits the framework-object to a definite entity. Thus in topology we might treat *space* either as a framework-object or as an entity. The mathematical treatment of the entity, space, is nonetheless mathematics—even if it is a special case. As we said before, the development of mathematics until quite recently has been from the point of view of the treatment of certain entities. This accounts for the fact that there is a confusion in the meaning of the word *space*, for example, even in mathematics. When we deal in mathematics with so-called abstract spaces we are really dealing with framework-objects.

Our concern here will be with the framework-objects. That there are different framework-objects is evidenced by the fact that we have various branches of mathematics. In what sense can we say that these objects exist? Are they real, logical constants, or merely fictions? Are they known empirically or by some form of direct apprehension?

We list some of the objects which occur in various branches of mathematics: *point*, *line*, *plane*, *projection*, *transformation*, *number*, *functions*, *matrix*, *group*, *space*.

These are, I believe, the most fundamental. All other concepts which occur can be defined in terms of these. Some of these may also be defined in terms of the others. Thus, an integral may be defined in terms of functions, and *space* may be defined as a group of points. Projections and transformations may be considered as types of functions. The existence (as we shall demonstrate later) of a one-to-one correspondence between real numbers and points on a line makes it possible that *number* as a framework-object has the same structure as *point* as a framework-object.

One other matter should be clear before we enter this discussion. We shall not be concerned here with the existence proof *in* mathematics. That is, we are not concerned with the methods of proving the existence in the mathematical structure of a new entity which has been defined.[2] We are rather concerned here with what the philosophers call the "ontological status" of the fundamental objects of mathematics.

[2] For a brief resumé of various positions on the idea of existence, cf. A. Fraenkel, "La Notion D'Existence," *L'Enseignement Mathématique*, Vol. 34, No. 1, 1935, pp. 18 ff.

Five characteristic positions have been taken with respect to the object of mathematics.[3]

1. The Intuitionist Position (old and new formulations).
2. Empirical Position.
3. Conventionalistic Position.
4. Logistic Position.
5. Formalistic Position.

OLD AND NEW INTUITIONISM

The word *intuition* has a number of meanings in contemporary discourse. It is used often to mean some form of mysterious experience. This meaning we rule out at once. The word in its original meaning has the significance of "experience." To intuit, then, is to experience. This makes it possible to add adjectives to the word and have different kinds of intuition. Thus, there is sensory intuition, rational intuition, and mystical intuition. Whether one can have pure sensory or pure rational intuition is not of concern here. The intuitionists in mathematical philosophy hold that pure rational intuition is possible and that the object of mathematics is essences, or entities, or ideas, related to man's faculty of knowledge and, therefore, known through reason and not through sense. Mathematics is concerned with objects which are not amenable to ordinary methods of observation.

We shall consider briefly two intuitionists, Kant and Brouwer. Brouwer wil' be discussed in great detail later since he is the founder of an entire approach to the foundations of mathematics.

Kant held that all sensory experience presupposed certain forms in which this experience took place. Of these forms, necessary for experience to be possible, there were two—space and time. Since these forms of experience were the "molds" in which we got experiences, it is not possible to know the forms through sensory experience. They are therefore known in *a priori* (i. e., prior to sensory experience) intuition. Mathematics is the system of those judgments in which the properties of the forms of our experiences are enunciated. These judgments are the result of *a priori* intuition and deduction. They are not the result of sensory intuition (i. e., they are not a posteriori).

Brouwer's intuitionism was used to form a foundational approach to mathematics. Brouwer insisted, as we shall see later, on the dependence upon human thought of all mathematics. The mathematical object as a consequence is likewise dependent upon thought. All concepts are known through rational intuition since they are actually

[3] Walter Dubislav, *Die Philosophie der Mathematik in der Gegenwart*, Berlin, 1932, p. 48 ff. I follow this closely. Cf. in connection with this discussion: Thomas Greenwood, "Invention and Description in Math."; *Arist. Soc. Proc.*, 1929–30, new series, Vol. 30, pp. 88 ff.

the result of rational processes. The so-called objects of mathematics are constructions based upon an original intuition of *twoness*.[4]

EMPIRICISM

In opposition to the intuitionist approach is the point of view that all knowledge comes from sensory experience. This does not deny the ability of the mind to abstract from its experiential data. It is through this process of abstraction that we arrive at the object of mathematics. Abstraction based upon our experiences of pairs of objects gives us "2." Mathematical objects are obtained by abstracting from color, shape, etc., until we arrive at the most abstract of abstract ideas. This type of approach leads very easily to the consideration of mathematics as a natural science. Later we shall discuss an empirical approach in the work of Pasch.

CONVENTIONALISM

A convention is an agreement to use, or do, something in a specific way. The purpose of a convention is usually one of convenience. Thus, the conventionalists insist that there is no mathematical object as such. Mathematics is a grandiose system of consistent conditions establishing certain conventions with respect to the symbols used. Thus, geometry is a system of consistent conditions agreeing as a matter of convention that lines shall be such that they intersect in only one point, if they are both in the same Euclidean plane. This position is close to formalism but it does not accept the assertion of the formalists that mathematics treats of symbols.

LOGICISM

The logistic school, which we shall also discuss in detail, insists that all mathematics is ultimately derivable from the concepts of logic. This is clearly evidenced in the definition of mathematics given by Russell and stated in the previous chapter. Thus, for the logistics the objects of mathematics are those constructs of logical concepts which satisfy certain conditions with respect to the variables they contain. Clearly, if that is the true object of mathematics then the logistical school faces the problem of demonstrating its assertion by an actual derivation of the concepts of mathematics from logic. To what extent this has been accomplished we shall discuss later.

FORMALISM

Formalism denies the necessity of the question from a purely technical point of view. The formalist defines his objects to be the symbols themselves *qua* symbols merely. He is interested in constructing sets

[4] Dubislav, *op. cit.*, p. 44.

of axioms for the symbols and to manipulate these symbols in prescribed ways. The formalist may or may not deny the existence of an object in any other sense. If he does not make such a denial, he does refuse to consider the question as of any relevance to him as a formal mathematician or logician. Mathematics is a calculus which may be interpreted, but this interpretation is unnecessary from a technical standpoint. The formal character of mathematics shows that the images associated with the terms in mathematics are not necessary to those terms.[5]

These are the most important positions relating to the objects of mathematics. They vary through almost all possible attitudes from the denial of any object at all to the assertion of a definite object. There are those who assert the existence of an object and identify it with "forms" of experience or with some definite property of objects. There are those who deny the existence of any object and attempt to identify the task of mathematics either with pure calculation or with convenient affirmations. These varying attitudes are reflected in the definitions given above. Number is treated either as a real independent existent (Plato) or as a pure symbol (Hilbert) or as a construct from logical concepts (Russell) or even as a pure fiction which is to be considered *as if* it were real (Vahinger).[6]

We have in Lewis' definition a formalist outlook, in Korselt's, Whitehead's, and Peirce's definition, logistical influences, and in Papperitz almost an empirical approach. However, few of these definitions belong clearly in one or the other of these positions. Sometimes in the writings of a single individual a number of these approaches are reflected. Present discussions, however, seem to be dominated to a great extent by syntactical problems. The various definitions as given present, to a great degree, an amount of ambiguity in their terms which makes impossible adequate discussion.

The definitions given actually do not contradict each other. They all seem to select some aspect either of pure or of applied mathematics and treat that as the essential characteristic. As a consequence, criticism of these definitions can take the form of demonstrating the breadth of the proposed definition. One could almost construct a definition of mathematics by stringing together the different terms in each. But it might be possible to approach the problem from a different angle.

That mathematics is a theoretical structure which attempts to give us grounded knowledge is evidenced in all definitions and in all systems

---

[5] Cf. Black, *op. cit.*, pp. 37–40.

[6] The best discussion of fictions with respect to Mathematics is in C. Betsch, *Fiktionen in der Mathematik*, Stuttgart, 1926.

of mathematics now known to us.[7] We do not have a branch of mathematics, i.e., a system of mathematics, unless and until we have obtained such a theoretical structure of grounded knowledge. By grounded knowledge we mean knowledge whose necessity is seen to flow from certain definite foundations. The form of this grounded knowledge is that of deduction—grounds to their consequences. The origin of our knowledge of that in which the system is grounded will determine whether the system is naturalistic or rational or arbitrary. In any case, given these grounds the rest must follow in deductive fashion. If we restrict our attention to this deductive aspect of mathematics, our definition of mathematics will contain such elements as implication, reasoning, constants, relations, and groups.

But a theoretical structure of grounded knowledge is a general form (a "Lehrbegriff"), i.e., a system. The structure of this system was indicated by Korselt's and Russell's definitions and will be treated separately later. As a general form it must be a form applicable to certain subject matters—as grounded knowledge it is *knowledge of something*.

Now, so far as we can see, there is no object knowledge of which cannot be put into this form. This does not mean that this form is the form into which *all* types of knowledge can be thrown; since otherwise, if the essence of mathematics were its form, all knowledge whatsoever (i.e., every science) would be potentially mathematics. Since not all sciences are mathematics, yet mathematics can give us knowledge about any object, it would appear that mathematics treats a certain *aspect* of any object. This means that mathematics would give us knowledge of that aspect of objects, i.e., would have as its object a definite aspect of objects.

Hilbert begins his classic *Foundations of Geometry* with the words "Let us consider three distinct systems of things"—one system of things he calls *points*, another *straight lines*, and the third *planes*.[8]

Lefschetz begins his work on topology[9] with the definition "Topology or Analysis Situs is usually defined as the study of properties of spaces or their configurations invariant under continuous transformations."

Mengenlehre deals with "multiplicities conceived as unities."[10]

All of these indicate that they are dealing with a definite type of object. In other words, the texts of mathematics bear witness to the fact that they do deal with a definite object. It would indeed be

[7] A great deal of the ideas expressed here was inspired by a reading of Husserl's, *Logische Untersuchungen*, Vol. 1, Chap. II. Many of the ideas will be evidently those of Husserl. Cf. also *Formale und transcendentale Logik*, Jahrbuch f. Phil. Vol. X, p. 67, und *Phänomenologische Forschung*.

[8] Cf. David Hilbert, *Foundations of Geometry*, 2nd ed. Chicago 1921, p. 3.

[9] Solomon Lefschetz, *Topology*, Am. Math. Soc. Coll. Pub. Vol. XII, 1930.

[10] Cf. Felix Hausdorff, *Mengenlehre*, Berlin u. Leipzig, 1927. p. 11.

strange if mathematics, although thought of as a science, would yet differ from all other sciences in the fact that it did not have a sphere of objectivity to which it referred.

It has frequently been pointed out that these names may be replaced by any others and all the theorems would still hold "true." Thus Hilbert's systems of points, lines, and planes, could be replaced by systems of x's, y's, and z's and all the theorems would still follow. This has usually been taken to mean that the terms were variables and since any entities could be put in the place of these variables, that mathematics had no object of its own. It appears, however, that another point of view is possible. If any entity can be put in the place of the terms used, then that seems to imply that mathematics is concerned with properties common to *all* these entities. These properties have been called the *quantitative* properties of objects. (The word "quantitative" has lost a great deal of its original clarity.)

The quantitative properties of objects are those properties of objects which they have in common. But the properties which all objects have in common can be only those properties which belong to them as *objects* merely. The particular type of object is of no concern. This means that mathematics treats those properties which things have in virtue of being objects; or to put the matter in another way, those properties which determine the possibility of objects. *The object of mathematics is then those properties of entities in virtue of which they are phenomenal (i.e., spatio-temporal) entities.*

Certain difficult problems are solved when we once recognize the object of mathematics. Since this object is the most general properties of entities as such, the various branches of mathematics, insofar as they are distinct, are seen to treat different types of such properties. The geometries treat the so-called extensional properties, arithmetic the ratios between properties, etc. The possibility of a mathematical treatment of any entities whatsoever is obvious. Mathematics and reality are not separate but one. Furthermore, no advance in mathematics can ever overthrow this description, since that advance will merely be a discovery of a new property of entities as such. Mathematics in itself can be defined as the theoretical treatment of these entities.[11] Hence we can see why some have defined mathematics merely as the structure itself without recognizing the nature of the mathematical object. The mathematical treatment of for example physics or astronomy, is merely a consideration of the properties of entities as they are found in physical or astronomical objects. Even Russell's definition views mathematics from the point of view of the theoretical structure. If we consider all the definitions given in Chapter Two and the list of concepts in those

[11] Cf. E. Husserl, *Ideas* (trans. by W. R. Boyce Gibson) 1931, pp. 65–66.

definitions, it will be evident that they all either involve elements of the theoretical structure or of properties of objects as such.

Insofar as we recognize the nature of the object of mathematics we can also see why an application of any mathematical theory is a matter of time. Since mathematics gives us properties of objects as such, it is theoretically possible to find, sooner or later, objects which clearly exemplify these properties. Interpretation of mathematical theories is therefore inevitable, although at any given moment of time not all the properties which objects as objects can have are clearly evidenced. But, as Leibniz has already pointed out, there is a unique relation between the object and the formulae of mathematics.[12] This is as true of the transcendental *Mengenlehre* as it is of Euclidean geometry. Existence proofs are seen to be necessitated by the fact that new mathematical objects (i.e., properties of objects as such) are introduced by means of *definitions.* It becomes necessary then to demonstrate that these new objects are possible objects and are capable of synthesis with objects already given. These new properties are for the most part not analytically derived but are introduced without any grounds. They must therefore be shown as capable of coexisting with the properties already present. Any new mathematical object derived by analysis needs no proof of existence. But where there is involved a free act of synthesis, the possibilities of this synthesis must be demonstrated. (This, of course, is a position already taken by Kant.)[13]

Such a position also makes evident the fact that Brouwer and the Intuitionists are striking in the right direction when they insist that the fundamental intuition of number is the basis of all mathematics. This is actually an intuition of a property of objects-as-such. Hilbert's insistence upon number as a *symbol* merely[14] is a mistake which resulted from the fact that the symbol was applicable to many objects and appeared to be the result of abstraction. These positions will be discussed in greater detail later.[15]

---

[12] Cf. E. Cassirer, *Philosophie d. symbolischen formen* Vol. 3, pp. 416–17.

[13] *Ibid.*, p. 420.

[14] Hilbert, *Neubegründung der Mathematik* (Abh. Math. Sem. Hamb. Univ.) 1922, p. 162. Also "Uber das Unendliche," *Math. Ann.* 95, 1926, pp. 170 ff.

[15] Cf. The excellent treatment of A. Heyting, *Mathematische Grundlagenforschungen*, Berlin, 1934, pp. 1–13, 57–62.

CHAPTER FOUR

A. FREGE'S DEFINITION OF NUMBER[1]

Gottlob Frege's attack on the foundations of arithmetic was in protest against the then current attempts to found arithmetic on psychological processes. He felt, justly, that a foundation for arithmetic could be a possible one only if it enabled us to develop the laws of arithmetic facts. A definition of number which is of no use to the mathematician cannot be considered a satisfactory definition. These ideas were developed in two works: *Die Grundlagen der Arithmetik* published in 1884, and *Grundgesetze der Arithmetik*, Volume 1 of which was published in 1893 and Volume 2 in 1903. (Frege was also of great importance in the history of logistic, in many phases of which he was Russell's and Whitehead's predecessor. Cf. *Begriffsschrift*, 1879.)

FOUNDATIONS OF ARITHMETIC

When we examine the structure of arithmetic we find that there are definite numerical equations involving specific numbers (such as $2 + 3 = 5$) and general laws which hold for all whole numbers. These must be clearly distinguished. Equalities (as $2 + 3 = 5$) have been held by some to be a sort of axiom—undemonstrable, and by others to be demonstrable. Actually if we consider these numerical equalities carefully, we find that they can be derived from the definitions of the individual numbers and certain general laws. Leibniz[2] recognized this to some degree when he tried to demonstrate that $2 + 2 = 4$ as follows:

Given: Def. (1) 2 is 1 and 1
(2) 3 is 2 and 1
(3) 4 is 3 and 1

Axiom: If we replace equals by equals in an equation, the equality continues to hold.

Proof: $2 + 2 = 2 + 1 + 1 = 3 + 1 = 4$
therefore, $2 + 2 = 4$.

---

[1] The reader will find translations of sections of the *Grundgesetze* in P. E. B. Jourdain and Johann Stachelroth, "The fundamental laws of arithmetic," *Monist*, Vol. 25 (1915), pp. 481–94; P. E. B. Jourdain and Johann Stachelroth, "The fundamental laws of arithmetic: psychological logic." *Ibid.*, Vol. 26 (1916), pp. 182–99; and P. E. B. Jourdain and Johann Stachelroth, "Class, function concept, relation." *Ibid.*, Vol. 27 (1917), pp. 114–27. Cf. also Bertrand Russell, *Principles of Mathematics*, 2nd ed., 1938, pp. 501–08.

[2] G. W. Leibniz, *New Essays Concerning Human Understanding* (trans. A. G. Langley), 1896, p. 472.

But Leibniz was using the general principle $a + (b + c) = (a + b) + c$.

Frege recognized what was later fully developed—*that the truths of arithmetic could be related to those of logic as the theorems of geometry are to the axioms of geometry.* Arithmetic would then be developed on the basis of definitions of the particular numbers and general propositions or laws which follow from the general concept of quantity (*Anzahl*) and which are best explained in terms of the *one* and *increase in ones*. Frege proceeds to attempt a derivation of number from logic as a demonstration of the truth of his thesis. In general, Frege works from the notion of number back to logical concepts.

Numbers, Frege points out, appear in language for the most part as adjectives and attributes in much the same way as do *hard*, *heavy*, *red*, etc., which denote properties of external objects. But numbers differ from these other attributes in that although the color of an object cannot be altered no matter what viewpoint is adopted, the object may be considered as being either (say) one poem, or twenty-four verses, or a great number of lines. Again, when we speak of green leaves, we mean each leaf is green, but if we speak of one hundred leaves, it cannot be said that each leaf is one hundred. The *number* of a group of objects depends upon the point of view we adopt in looking at the objects. This is obviously not the case with color, weight, etc., of an object. Therefore, although number appears as an attribute, it is not an attribute which is obtained by abstraction from external objects. Number is a property but not a property of single objects.

But we cannot conclude from this that number is a purely subjective entity, as some would do. For just as it does not follow that the North Sea is a subjective entity because the part we look at depends on our choice, so it does not follow that number is other than objective. Quantitative predication is independent of our mental images. For example, the equator is objective but not in the same sense as the earth. By objectivity is meant independence of our sensations, intuitions, and images, but not independence of reason. Arithmetic is not psychology. (Cf. *Grundgesetze*, Vol. 1, p. xviii.)

Frege means that number is an objective aspect of a plurality. The number of a plurality may be altered by changing one's point of view; but, from a given point of view, it is possible to predicate only one number. This notion of number as an objective property is later rejected by the formalists in favor of a more nominal approach. Number is considered by them as an arbitrary symbol and only the demands of consistency prevent us from predicating different numbers of the same group.

When I look at some trees and say "this is a group of trees," or when I look at the same trees, am more specific, and say "there are five trees"; or, to take another illustration, when I look at soldiers and say "there are four companies," or "there are five hundred men"—then the group of trees or of soldiers has not been altered. All I have changed by being more specific is the name of the group, not the group or its extension. That is, I have replaced one concept by another. (Frege uses the term *concept* where we would today use the term *class*.) A class is defined by a propositional function with one variable; hence the term *concept* here means also nearly the same thing as propositional function.[3]

Thus, to make an assertion about a concept really means to make an assertion about a class, or to predicate something of a propositional function. Number then is a property of concepts. But this does not make it subjective. If I say "all men are animals," this proposition is one about the concepts "men" and "animals." Obviously the concept "man" is not completely subjective, even though it does not refer to any particular man. It is only when we can apply the concept to one or more objects that we have the possibility of number. The objects to which the concept can be applied may be said to be subsumed under it. These objects then constitute the *extension* of the concept. Since a concept is a predicate, the extension of the concept (or class) consists of those objects which are subjects for the predicate. If the predicate is such that it cannot be applied to any objects, the extension is said to be *null*. Numbers are predicates of concepts denoting something about the extension of the concept. The idea now is to begin with a concept having a null extension, of which 0 is then predicated, and to build up the natural number series.

If no object is subsumed under a given conception, i.e., if no object satisfies the defining properties of the concept, then we attribute the number 0 to that concept. This means that 0 is predicated of a concept, if no matter what a is, the proposition "a is not in the extension of that concept," is a valid proposition. Or to say the same thing in other words, 0 is predicated of a class if for every entity a, the proposition "a is not an element of the class," is always valid. Likewise 1 is predicated of a concept F if no matter what a is, the proposition "a is not an element under F" is not in general a valid proposition, and if from "a is an element of F" and "b is an element of F," it always follows that "a is the same as b." There remains the task of defining a method of generating from any number its successor. We attempt this as follows:

The number $(n + 1)$ is attributed to the concept F if there exists an object a which is subsumed under F and if F is so constructed that the number n is attributed to the concept "subsumed under F but not equal to a." This means that a concept F has the number $(n + 1)$ if the

[3] Cf. Bertrand Russell, *op. cit.*, p. 507.

concept which remains after an element is subtracted from F has the number n.

Actually we have not defined the number 0 or the number 1, but merely:

"the number 0 is predicated of"
"the number 1 is predicated of."

We have no means here of distinguishing 0 and 1 from these. It is necessary to define certain relations which occur in arithmetic, as, e.g., equality.

Since 0 and 1 occur in propositions, their significance will depend upon these propositions. Since, further, number-words denote self-subsistent entities, we consider only those propositions which express a recognition of such objects. If the symbol a denotes an object, then we must have some means of determining in every case (at least theoretically) whether or not b is the same as a. It is necessary for us to explain the proposition:

> The number which is predicated of concept F is the same as the number which is predicated of concept G.

In this way we would define a general condition for equality of numbers. This condition is found in the idea of one-to-one correspondence, i.e., if each unit of one concept corresponds to only one unit of the other, and conversely, then the numbers are said to be the same. Hence two concepts F and G are said to be of *equal number* if a one-to-one correspondence between them can be set up. This definition may be stated more precisely as follows:

> The concept F is said to be of *equal number* with concept G, if there exists a relation ϕ, which correlates in a univocal (1,1) correspondence the objects falling under F with those falling under G.

We can take the following example. Consider the concept F "man who is in this room" and the concept G "chair which is in this room." If there is but one man for each chair and only one chair for each man, then we say there is a one-to-one correspondence between F and G, and the numbers of F and of G are the same.

Here by a univocal (1,1) correspondence Frege means a correspondence such that the following two propositions are valid:

1. If d has the relation ϕ to a and if d has the relation ϕ to e, then a and e are equal, no matter what a, d, and e are.
2. If d has the relation ϕ to a, and b has the relation ϕ to a, then no matter what a, b, and d are, b and d are equal. But this really means that one can be substituted for the other. We define in accordance with these remarks:

> The quantity which is predicated of concept F is the *extension* of the concept "of equal number with concept F."

This says that the quantity which is predicated of a concept F is the class of all classes which can be put into one-to-one correspondence with each other. Thus we have explained the cardinal number of a concept in terms of the extension of concepts (i.e., classes) whose extensions can be put into one-to-one correspondence with each other. It is necessary to show that the properties of number follow from the definition before we can accept the definition. What Frege now does is to define 0, then to define "next number in the natural number series," from which he gets 1. From 1 he gets the number 2 by constructing the next number to 1, and so on.

Since the concept "not equal to itself" has no extension we define:

> 0 is the quantity which is predicated of the concept "not equal to itself."

To say that a occurs under the concept "not equal to itself" is to say that a is not equal to itself. This is clearly false, no matter what a is. Further, in order to define the "next" element in the number series we have the following:

> n follows m immediately in the natural number series, if there exists a concept F and an object x subsumed under F, such that the quantity predicated of the concept F is n and the quantity predicated of the concept "subsumed under F and not identical with x" is m.

Combined with the definition given above, n is actually $(m + 1)$.

With the use of this definition, we can proceed to define the number 1 as the number which follows 0 immediately in the natural number series. This means that it is necessary to show that there exists a concept F having an element subsumed under it such that the quantity of the concept "falling under F but not equal to x" is 0. This latter class must therefore be a null class. Since we have defined 0, we can make use of this number 0. If, now, we let F be the concept "equals 0," then clearly the 0 is an element of this class. Furthermore, the concept "equals 0 but not equal to 0" is a null concept since there can be no element subsumed under it. It therefore follows that the quantity of F follows the 0 immediately in the natural number series.

We can now define the 1:

> 1 is the quantity predicated of the concept "equal 0"; i.e., 1 is the immediate successor of 0 in the natural number series.

(Frege insists that no observed object was involved in this definition of *one*. Number thus, for him, would appear to be definitely demonstrated not to be an abstraction. But this need not be true, as our discussion of Pasch later will show.)

Having defined 0 and 1, there remains the task of demonstrating that the natural number series is unending. This will be done if we can show that every number n in the natural number series has an immediate successor in that series. In accordance with the procedure used in defining the number 1 the theorem will be demonstrated if we can define a concept of which we can predicate this latter number. In other words we must show that there exists a concept whose number is $n + 1$ for any n we may select. We can select as this concept:

"belongs to the natural number series ending with *n*."

We must show that the number predicated of this concept follows *n* in the natural number series. To do this it is necessary to define the expression "follows in a series." Thus:

"*y* follows *x* (or *x* precedes *y*) in the ϕ-series" means: "*If*

1. Any object to which *x* has the relation ϕ is subsumed under the concept *F*, and
2. From the fact that *d* is subsumed under *F* it follows that, no matter what *d* is, any object to which *d* has the relation ϕ falls under *F*

then y falls under *F* no matter what *F* is."

This definition, as Frege points out, enables us to refer mathematical induction to the general laws of logic.

This is an almost literal translation of the definition as given by Frege. In its present form it is not very clear. Apparently what is meant is this: "*y* follows *x* in the ϕ-series" means that you can start with *x* which is subsumed under *F* and a relation ϕ such that every object to which *x* has the relation ϕ is also subsumed under *F* and you can proceed from *x* to any element *d* which has the relation ϕ to *x* and is therefore also subsumed under *F*, and from *d* to the element *y* which has relation ϕ to *d* and is therefore subsumed under *F*. As an example consider the positive whole numbers. Then 5 is said to follow 2 in this series since you can take the relation "is less than by one" and go from 2 to 3, from 3 to 4, and from 4 to 5 using this relation. The relation "is less than by 1" is moreover such that anything to which a number predicated of *F* bears this relation must also be a number.

If now the relation ϕ between *m* and *n* is that which is defined by the proposition:

"*n* follows *m* immediately in the natural number series," then in place of ϕ-series we say 'natural number series'."

One other definition is needed:

"*y* belongs to the ϕ-series beginning with *x*" means that "*y* follows *x* in the ϕ-series or *y* is the same as *x*."

The proof that there is a natural number after any natural number consists in showing that—under a condition to be defined—the quantity predicated of the concept "belonging to the natural number series ending with *n*" follows the quantity *n* immediately in the natural number series. Since the series ending with *n* can be enlarged always to include the last quantity defined, there can be no last term to the natural number series.

We shall merely indicate the proof. Two propositions are to be demonstrated:

1. If a follows d immediately in the natural number series and d is such that the quantity predicated of the concept "belonging to the natural number series ending with d" follows d immediately in the natural number series, *then* it is also true that the quantity predicated of the concept "belonging to the natural number series ending with a," follows a immediately in the natural number series.

2. The above proposition is satisfied by 0 and therefore it also holds of n if n belongs to the natural number series beginning with 0.

Now, in order to prove 1, we must show that a is the quantity predicated of the concept "belonging to the natural number series ending with a but not equal to a." And for this we need to show that this concept has the same extension as the concept "belonging to the natural number series ending with d." This needs the proposition that no object which belongs to the natural number series beginning with 0 can follow itself in the series—which must be demonstrated by means of the definition of "follows in a series."

In these analyses of Frege certain features are predominant and should be kept in mind. The entire derivation of arithmetic (from which, of course, other branches of mathematics follow) is based upon "concept," "belonging to the extension of," and "predication." These are logical concepts and hence arithmetic seems to follow from logic. Frege himself says, "arithmetic is only an extended logic, every arithmetic proposition is a law of logic derived from logic." No observation of empirical objects appears to be necessary and the laws of arithmetic are apparently not natural laws (unless indeed laws of logic are natural). The ultimate nature of number and mathematics in general can on this theory be decided only after we know the nature of the laws of logic. If the laws of logic are purely arbitrary, then mathematics is also. In any case, mathematics deals with objects, concepts, and relations as does logic. Furthermore, number is here considered as a predicate of "concepts," i.e., of classes—of objects as forming groups. Number therefore pertains to that aspect of any object in virtue of which the object can form a group with other objects. That this implies the ability of the mind to form groups, Frege does not specifically recognize. Pasch's approach which we study later indicates another angle to the problem.

This describes the general approach which has dominated, with modification, the work of the Russell-Whitehead school and to some extent the work of others. The technical detailed construction of the logistic system as developed by Frege is based upon a symbolism which is difficult, cumbersome, and frightening. The historical importance of the work cannot be overestimated and has been the subject of much investigation by H. Scholz. We, therefore, place the discussion of the system later as an appendix to the Russell-Whitehead chapter.[4]

B. RUSSELL AND THE LOGISTIC DEFINITION OF NUMBER

Since for the Logistic foundations of mathematics, mathematics is a branch of logic, the logisticians have certain very definite problems confronting them: (1) Every mathematical concept must be demonstrated to be a logical concept; (2) All methods of proof used in mathematics must be shown to be such as are purely logical methods; (3) Every mathematical proposition must be deducible from the fundamental assumptions of pure logic alone.[5] Where mathematics begins and logic ends is a purely arbitrary division. Actually they are one. Mathematical propositions are therefore, for logistics, propositions of logic, i.e., purely analytic *a priori* propositions, and mathematical concepts are combinations of the fundamental concepts of logic. A proposition is said to be analytic if the predicate is but an explication of the subject; it is *a priori* if it is not derived from experience.

Frege, as we have seen, had already started to solve the first problem. Russell introduces this same problem in his *Principles of Mathematics* (First published in 1903 and reprinted in 1937) in nonsymbolic form and later in the three-volume work with A. N. Whitehead, *Principia Mathematica* (Vol. 1 published in 1910, 2nd edition 1925; Vol. 2 in 1912, 2nd edition 1927; Vol. 3 in 1913, 2nd edition 1927. The second edition was reprinted in 1935. The symbolism which Whitehead-Russell adopts has its roots in the work of the Italian Peano and his collaborators.)[6] The new symbolism which Whitehead-Russell (hereafter to be cited as *P.M.*) uses makes possible the analysis of the fundamental concepts and methods of proof involved in mathematics more clearly and precisely. This symbolism is now the most commonly used one.[7] The first and second problems above necessitated a careful extension of both the logic of concepts and the logic of deduction, which *P.M.* develops in great detail. The third problem of the logistician seems to be unsolved since it apparently involves the introduction of two axioms: (1) axiom of infinity, and (2) multiplicative axiom, which do not lend themselves to derivation from pure logic, i.e., they are not

[4] Cf. the recent article by H. R. Smart, "Frege's Logic," *The Philosophical Review*, Vol. 54(1945), pp. 489–505. Also the review of this article by A. Church in the *Journal of Symbolic Logic*, Vol. 10(1945), pp. 101–03.

[5] Cf. W. Dubislav, *Die Philosophie der Mathematik in der Gegenwart*, 1932, pp. 38–39.

[6] Cf. G. Peano, *Formulaire de Mathématiques*, Vol. 1, Torino, 1895.

[7] The reader will find Bertrand Russell's *Introduction to Mathematical Philosophy*, 1919, Chap. 2, of value. Cf. also Max Black, *The Nature of Mathematics*, London, 1933.

a priori analytic propositions.[8] However, these axioms are needed only when we wish to extend our analyses to include problems about the continuum and transfinite numbers.

In order to develop the logistic definition of number we shall follow *P.M.* in outline, omitting important logical doctrines and detail.[9] The sections we are interested in occur in Volumes 1 and 2 and they are Part I, Mathematical Logic, Part II, Prolegomena to Cardinal Arithmetic, and Part III, Cardinal Arithmetic. There is no difference in the numbering of the propositions in the two editions of *P.M.* We are using the second edition, which includes an introduction of value for those interested in the development of symbolic logic. (The modifications suggested by the work of Nicod and given in Appendix A of *P.M.* are not used because they have not generally been adopted.)

PART I

Elementary propositions, i.e., propositions which do not contain variables, will be denoted by $p, q, r, s, \ldots$ For example, "this is red" is an elementary proposition. Combinations of elementary propositions are also elementary propositions. For example, "this is red" and "Socrates is mortal" are both elementary propositions, as is also the combination "this is red and Socrates is mortal." $\vdash p$ will denote that p is asserted. p by itself will denote the proposition p. $\vdash p$ may be read "it is asserted that" or "it is true that." $\sim p$ will denote the falsity of p (read "not-p"). If p and q are any propositions, then *p or q*, i.e., either p is true or q is true, will be denoted by $p \vee q$. Also $\sim(p \vee q)$ will denote that it is false that either p is true or q is true, i.e., both are false. $p \vee q$ is called the *disjunction* or *logical sum* of p and q.

1.01 $p \supset q . = . \sim p \vee q$ Df.,[10]

i.e., "p implies q" means that either p is false or q is true. Thus the truth or falsity of $p \supset q$—i.e., the truth-value of $p \supset q$—depends upon the truth-values of p and q. The letters Df. mean that this is a definition. The horseshoe symbol stands for the concept "implies," and is often read "if . . . then . . . ." The dot represents the conjunction or logical

[8] The axiom of infinity postulates the existence of infinitely many *propositional functions*. The multiplicative axiom (also called Zermelo axiom and Axiom of Choice) states "Given any class of mutually exclusive classes, of which none is null, there is at least one class which has exactly one term in common with each of the given classes." A. Fraenkel, *Einleitung in die Mengenlehre*, 3rd ed. 1928, and J. Jörgenson, *Treatise of Formal Logic*, 3 vols., 1931, have good discussions of these axioms.

[9] Cf. also Rudolf Carnap, *Abriss der Logistik*, Vienna, 1929. Excellent introductory texts are J. C. Cooley, *Primer of Formal Logic*, 1942; S. Langer, *Introduction to Symbolic Logic*, 1937; A. Tarski, *Introduction to Logic*, 1941.

[10] The numbers are those in *P.M.* The student would do well to read Chapter II of Russell's *Introduction to Mathematical Philosophy* before tackling this analysis. The symbol $\ldots = \ldots_{Df.}$ means "is the same as, by definition."

product. It is read "and." Where the single dot is not clearly conjunction, and where more than one dot occurs, parentheses are indicated.

3.01 $p.q. = .\sim(\sim p \vee \sim q)$ Df.

4.01 $p \equiv q. = .p \supset q.q \supset p$ Df.

i.e., if two propositions mutually imply one another they are said to be identical, i.e., they have the same truth value.

4.1 $\vdash : p \supset q. \equiv .\sim q \supset \sim p$

4.13 $\vdash .p . \equiv . \sim(\sim p)$

4.3 $\vdash : p.q. \equiv .q.p$

4.31 $\vdash : p \vee q. \equiv .q \vee p$

If we have an elementary proposition and replace the subject by a variable x, we get an elementary function. For example, if P represents the predicate of a, then Pa can be read "a is a P." If we now replace a by x and write Px, we have "x is a P," which is an elementary function. We get different values for Px as we substitute individual objects for x.

If we have the proposition that ϕx is true for all values of x we shall denote it by $(x)\phi x$. Then $(x):\phi x. \supset .\psi x$ will mean "for all x, ϕx implies ψx," i.e., all objects having the property ϕ also have the property ψ. To denote "ϕx sometimes" we shall write $(\exists x).\phi x$—read "there exists an x such that ϕx."

We now put

9.01 $\sim\{(x).\phi x\}. = .(\exists x).\sim\phi x$ Df.

i.e., "It is false that for all x, ϕx" means that there exists an x such that ϕx is false.

9.02 $\sim\{(\exists x).\phi x\}. = .(x).\sim\phi x$ Df.

10.02 $\phi x \supset_x \psi x. = .(x).\phi x \supset \psi x$ Df.

10.03 $\phi x \equiv_x \psi x. = .(x).\phi x \equiv \psi x$ Df.

10.1 $\vdash :(x).\phi x. \supset .\phi y$

i.e., what is true of all is true of any one.

10.24 $\vdash : \phi y. \supset .(\exists x).\phi x,$

i.e., if ϕy is true then there is an x such that ϕx is true.
We now let $\phi ! x$ represent any value of an elementary function.

13.01 $x = y . = : (\phi) : \phi ! x . \supset . \phi ! y$ Df.

i.e., "x is identical with y" means that no matter what function is used the *value* resulting for ϕx implies the *value* resulting for ϕy.
We now denote "the x which satisfies ϕx" by (ιx) (ϕx)

14.02 $E!(\iota x)(\phi x) . = : (\exists b) : \phi x . \equiv_x . x = b$ Df.

(read: "*the* x satisfying ϕx exists" means there exists a b such that for all x, ϕx implies and is implied by "x is identical with b.") We shall let $\hat{z}$ (ϕz) denote "those values of z which satisfy ϕ," i.e., it denotes the class defined by ϕz. "$x \epsilon \hat{z}(\phi z)$" means x is a member of the class determined by ϕz.

20.15 $\vdash : . \psi x . \equiv_x \phi x : \equiv . \hat{z}(\psi z) = \hat{z}(\phi z)$

i.e., if ψx is identical with ϕx for all x, then the class determined by ψx is equal to that determined by ϕx.

20.31 $\vdash : . \hat{z}(\psi z) = \hat{z}(\phi z) . \equiv : x \epsilon \hat{z}(\psi z) . \equiv_x . x \epsilon \hat{z}(\phi z)$

i.e., the class defined by ψz is equal to that defined by ϕz if and only if for every x, x is a member of $\hat{z}(\psi z)$ is identical with x is a member of $\hat{z}(\phi z)$.

20.32 $\vdash : x \epsilon \hat{z}(\psi z) . \equiv . \psi x$

i.e., x is a member of the class $\hat{z}(\psi z)$ if it satisfies the function which determines the class.

20.33 $\vdash : . \alpha = \hat{z}(\phi z) . \equiv : x \epsilon \alpha . \equiv_x . \phi x$

This enables us to use Greek letters for classes.

20.34 $\vdash : . x = y . \equiv : x \epsilon \alpha . \supset_\alpha . y \epsilon \alpha$

The inclusion of one class in another will be denoted by $\subset$.

22.01 $\alpha \subset \beta . = : x \epsilon \alpha . \supset_x . x \epsilon \beta$ Df.

i.e., "the class α is included in (or is a subclass of) class β" means that for all x, x is an element of α implies it is also an element of β, i.e., the elements of α are all elements of β.

22.02 $\alpha \cap \beta = \hat{x}(x \epsilon \alpha . x \epsilon \beta)$ Df.

i.e., the class denoted by $\alpha \cap \beta$ is composed of those elements of x which are elements both of α and of β at the same time—(read "α and β." This is the logical product)

22.03 $\alpha \cup \beta = \hat{x}(x \epsilon \alpha . \vee . x \epsilon \beta)$ Df.

i.e., the class of all elements either elements of α or of β—(read "α or β." This is the logical sum).

22.04 $-\alpha . = . \hat{x}(x \sim \epsilon \alpha)$ Df.

24.01 $V . = . \hat{x}(x = x)$ Df.

i.e., V is the class of those elements which are identical to themselves. Since any element whatsoever satisfies this description, V is the class which contains everything and is called the *universal class.*

24.02 $\Lambda = -V$ Df. By 22.04 $-V = \hat{x}(x \sim \epsilon V)$

i.e., $-V$ is class of those elements not members of V. Since V contains everything, there are no such elements, and is called the "null class."

24.03 $\exists ! \alpha . = . (\exists x) . x \epsilon \alpha$ Df.

i.e., there exists an element x such that $x\epsilon\alpha$, i.e., $\exists ! \alpha$ means α is not null.

A *relation* will be regarded as the class of couples (x, y) for which some function $\psi(x, y)$ is true.

21.33 $\vdash :. R = \hat{x}\hat{y}\phi(x,y) . \equiv : xRy . \equiv_{x,y} . \phi(x, y)$

i.e., "R is identical with the class of couples determined by $\phi(x,y)$" means that x is related to y implies and is implied by $\phi(x,y)$ for all x's and y's.

21.43 $\vdash :. R = S . \equiv : xRy . \equiv_{x,y} . xSy$

Further we shall define $R\text{‘}y$ (read "R of y") as the class of those x's which have the relation R to y, i.e.,

30.01 $R\text{‘}y = (\imath x)(xRy)$ Df.

Given any relation R, then we define:

32.01 $\overrightarrow{R} = \hat{\alpha}\hat{y}\{\alpha = \hat{x}(xRy)\}$ Df.

i.e., $\overrightarrow{R}$ is the relation of the class of referents of y to y, i.e., of all the class of terms which have the relation R to a given term y. Also

32.02 $\overleftarrow{R} = \hat{\beta}\hat{x}\{\beta = \hat{y}(xRy)\}$ Df.

i.e., $\overleftarrow{R}$ is relation of the class of *relata* of x to x, i.e., of all the class of terms to which a given term x has the relation R.

We now have:

32.13 $\vdash . \overrightarrow{R}\text{‘}y = \hat{x}(xRy)$

i.e., the referent of y is the class of x's such that xRy.

32.131 $\vdash . \overleftarrow{R}\text{‘}x = \hat{y}(xRy)$

From 32.13 and 20.33 it follows

32.18 $\vdash : x \epsilon \overrightarrow{R}\text{‘}y . \equiv . xRy$ and

32.181 $\vdash : y \epsilon \overleftarrow{R}\text{‘}x . \equiv . xRy$

We have now defined a relation and we can therefore set up correspondences. It is evident how closely this definition of number which we are developing follows that of Frege given in the previous section. Since similarity (cardinal, at least) is in terms of the elements defined by the relations and these elements constitute two classes, these two classes must be defined more precisely.

We therefore need:

33.11 $\vdash . D`R = \hat{x}\{(\exists y) . xRy\}$

i.e., $D`R$ (called the *domain* of the relation R) is that class of x's for which there is a y to which they have the relation R.

33.111 $\vdash . \text{ɑ}`R = \hat{y}\{(\exists x) . xRy\}$

$\text{ɑ}`R$ is called the *converse domain* of R.

For example,

> if R is the relation of "begetting," then
> $D`R$ is the class of all "fathers" and
> $\text{ɑ}`R$ is the class of all "offspring."

A relation R is said to be *one-one* if for any member y of the converse domain there is one and only one member x of its domain which has the relation R to y and conversely if for any member x of its domain there is one and only one member y of its converse domain to which x has the relation R. One-one relations will be said to be relations belonging to the class of one-one relations denoted by $(1 \rightarrow 1)$.

Two classes, α and β, are said to be *similar* if there is a one-one relation whose domain is α and whose converse domain is β.

Thus:

73.02 $sm = \hat{\alpha}\hat{\beta}(\exists ! \alpha \overline{sm} \beta)$ Df.

where

73.01 $\alpha \overline{sm} \beta = (1 \rightarrow 1) \cap \overleftarrow{D}`\alpha \cap \overleftarrow{\text{ɑ}}`\beta$ Df.

i.e., $\alpha \overline{sm} \beta$ is the class of relations which correlate α's and β's so that α is similar to β.

73.1 $\vdash : \alpha \; sm \; \beta . \equiv . (\exists R) . R \epsilon 1 \rightarrow 1 . \alpha = D`R . \beta = \text{ɑ}`R$

i.e., "α is similar to β" means there is a one-one relation R such that α is the domain of R and β is the converse domain of R.

We are now ready to define the cardinal number of a class α. This cardinal number, denoted by $Nc`\alpha$, is defined as the class of all classes similar to $\alpha$, i.e., as $\hat{\beta}(\beta \; sm \; \alpha)$. Since $\overrightarrow{sm}`\alpha = \hat{\beta}(\beta \; sm \; \alpha)$, $Nc`\alpha = \overrightarrow{sm}`\alpha$ and $Nc = \overrightarrow{sm}$

100.01 $Nc = \overrightarrow{sm}$ Df.

100.1 $\vdash . Nc\text{‘}\alpha = \hat{\beta}(\beta \; sm \; \alpha) = \hat{\beta}(\alpha \; sm \; \beta)$

100.11 $\vdash . Nc\text{‘}\alpha = \hat{\beta}\{(\exists R) . R \epsilon 1 \rightarrow 1 . D\text{‘}R = \alpha . \text{Ɑ}\text{‘}R = \beta\}$

i.e., the cardinal number of α is the class of all β such that there exists a one-one relation R of which α is the domain and β is the converse domain.

The procedure of the next part will be to define unit classes and couples and then 1 and 2. These are then demonstrated to be cardinal numbers.

We first define the *relation of identity*.

50.01 $I = \hat{x}\hat{y}(x = y)$ Df.

Hence $\overrightarrow{I}\text{‘}x = \hat{y}(y = x)$, i.e., $\overrightarrow{I}\text{‘}x$ is the class of terms identical with x, we shall write

51.01 $\iota = \overrightarrow{I}$ Df.

We therefore have

51.11 $\vdash . \iota\text{‘}x = \hat{y}(y = x)$

i.e., x is the class which has only x as element—called the unit class.

The cardinal number 0 is defined as:

54.01 $0 = \iota\text{‘}\Lambda$ Df.

i.e., 0 is the class of all classes which are null.

And we can now define the cardinal number 1.

52.01 $1 = \hat{\alpha}\{(\exists x) . \alpha = \iota\text{‘}x\}$ Df.

i.e., 1 is the class of those classes α such that there exists an x which is the only member of α

We add

52.13 $\vdash . 1 = D\text{‘}\iota$

i.e., 1 is the domain of the relation defining unit classes.

52.18 $\vdash :. \alpha \epsilon 1 . \equiv : (\exists x) : x \epsilon \alpha : y \epsilon \alpha . \supset_y . y = x$

i.e., a class is a unit class (i.e., has cardinal number 1) if when there is an x, and x and y are elements of α, then $y = x$, for all y.

It will be noticed that Russell's 1 is defined differently from Frege's. For Frege 1 is the class which has as its only member the null class; for Russell 1 is the class whose only member is the unit class. Russell apparently makes the unit class the fundamental entity of number. Both, however, make as the root of their definitions "identity"; $x = y$ or $x \neq y$.

We define 2 as follows:

54.02 $2=\hat{\alpha}\{(\exists x,y).x\neq y.\alpha=\iota`x\cup\iota`y\}$ Df.

i.e., 2 is the class of classes α such that there is an x and a y not equal to each other and α is the disjunction of the class whose only member is x and the class whose only member is y.

The theorem from which follows the demonstration that 0 and 1 are the only numbers less than 2 is:

54.4 $\vdash:\beta\subset\iota`x\cup\iota`y\equiv:\beta=\Lambda.\vee.\beta=\iota`x.\vee.\beta=\iota`y.\vee.\beta=\iota`x\cup\iota`y$

which says that if a class is contained in a couple it is either the null class or one of the unit classes or the couple itself.

It can be demonstrated that

73.45 $\vdash.1=\hat{\beta}(\beta\ sm\ \iota`x)$

i.e., 1 is the class of classes similar to any unit class.
Also

73.48 $\vdash.0=\hat{\beta}(\beta\ sm\ \Lambda)$

It clearly follows from 73.48 that 0 is the cardinal of Λ, i.e.,

101.1 $\vdash.0=Nc`\Lambda$

also

101.2 $\vdash.1=Nc`\iota`x$

i.e., 1 is the cardinal of any unit class, and

101.3 $\vdash:x\neq y.\supset.2=Nc`(\iota`x\cup\iota`y)$.

This is as far as we need go for our purposes. *P.M.* proceeds to define ordinal numbers, the arithmetic operations, series, convergence of series, continuity of concepts, etc.[11]

In this definition of cardinal number we find the following concepts:

class (α, β, ...)	element of a class (ϵ)
extension of a class,	one-one relation ($1\rightarrow1$)
relation (R)	domain of a relation ($D`R$)
existence ($\exists$)	converse domain ($\text{Ɑ}`R$)

Since relation, domain, and converse domain are reducible to classes, the list reduces to:

class	element of a class
extension	existence

[11] For a summary of this, the reader is referred to J. Jörgenson, *Treatise*, Vol. II, Chap. 8, and R. Carnap, "Die Logizistische Grundlegung der Mathematik," *Erkenntnis*, Vol. 2, 1931.

These are, however, the primitive terms of logical analysis. It would appear to follow that number is derivable from these terms, i.e., is itself a logical construct or property of classes.

In conclusion we might summarize: The logistic thesis concerning mathematics can be divided into two parts: (1) Mathematical *concepts* are derivable from logical concepts by means of explicit definitions. (2) Mathematical *propositions* are derivable from fundamental logical propositions by means of pure logical deduction.[12]

1. The logical concepts needed are: negation of a proposition, "not-p" (denoted by $\sim p$); disjunction of two propositions, "p or q" ($p \vee q$); conjunction, "p and q" ($p . q$); implication, "if p then q" ($p \supset q$). Also, concepts as such are denoted by functions, e.g., $f(a)$ (read "f of a") which denotes that a has the property f. Also, $(x) f(x)$ (read "for all x, f of x is true"); $(\exists x) f(x)$ (read "there exists an x for which f of x is true"); and finally, $a = b$, which means a and b are names of the same object. The essential thing about the solution of the logistic problem is the fact that the natural numbers are logical determinations which pertain not to objects but to concepts. That the quantity (say) 3, is predicated of a concept (in symbols, $3(f)$), means that three objects are subsumed under this concept. We can denote this by means of the concepts named above. For example, let $2m(f)$ mean that "at least two objects are subsumed under f."
Let $=_{Df.}$ be the symbol which denotes a definition:

$$2m(f) =_{Df.} (\exists x)(\exists y)[\sim(x=y) . f(x) . f(y)]$$

(read $2m(f)$ is defined as "there exists an x and there exists a y such that x is not the same as y, and f is predicated of x, and f is predicated of y.") In like manner we can define $3m(f)$.

Now the quantity 2 can be defined as

$$2(f) =_{Df.} 2m(f) \sim 3m(f).$$

We can from this point on define all types of numbers even though certain difficulties are involved in the definition of the real irrational numbers.

2. The necessary system of fundamental propositions of logic for the deduction of mathematical propositions results from a simplification of Russell's system. It contains four fundamental propositions of the logic of propositions and two of the logic of functions.

Mathematical concepts are introduced in terms of explicit definitions, i.e., as abbreviations of complexes of logical symbols. This

---

[12] R. Carnap, *op. cit.*, pp. 92 ff. What follows is part of this article.

means that every demonstrable mathematical proposition is translatable into a proposition containing only logical symbols, and demonstrable in logic.

APPENDIX: FREGE'S SYMBOLISM

Fundamental Laws of Arithmetic:[13] In these two volumes Frege develops the position outlined above in purely symbolic form. He feels and explicitly states that all propositions which are not demonstrated but are used should be expressed precisely, while all demonstrated propositions are to be derived as analytic propositions in a series of formulae (Vol. I, pp. v–vii, also p. 1). In order to do this, Frege introduces a very complex and cumbersome symbolism which has been superseded today by that of Russell and Whitehead described in the preceding pages. This symbolism has for its purpose the derivation of theorems so that no gap is left in any proof. This demand for extreme rigor is an outstanding characteristic of contemporary mathematics and has led to very significant developments in many branches of mathematics. Furthermore, this symbolism enables us to explore the complete development of arithmetic from logic in all its parts, so as to leave no doubt as to the fact that arithmetic is extended logic (p. vii). The difficulty of Frege's symbolism and novelty of his approach resulted for a time in the neglect of his work until he was rediscovered by Russell. [Frege himself bemoans the lack of attention paid by investigators to his work (p. xi).]

Frege believes his theory to be much more general than a purely logical theory. We shall restrict our discussion to the logical aspects only. These ideas will recur from now on in the work of every logician.

We define the expression

"the function $\phi(\xi)$ has the same *truth-value range* as $\psi(\xi)$"

to mean

"the function $\phi(\xi)$ and $\psi(\xi)$ have the same values (i.e., are both true or both false) for the same arguments (i.e., when the same constant replaces ξ in both functions)."

The truth values are *true* and *false*. ξ represents a variable whose *argument* is a definite range of constants. For example, given $\xi^2=4$. Then ξ is the variable, its arguments are 0, 1, 2, 3, 4,. . . . When $\xi=2$ the value of $\xi^2=4$ is *truth*, for all $\xi \neq 2$ it is *falsity*. Also if we have $\xi^2=4$ and $3.\xi^2=12$, then these two functions have the same truth value range, i.e., are both true for $\xi=2$ and false for $\xi=1, 3, 4,\ldots$. Where we have an equality such as $2+3=5$ we are not concerned from this

[13] G. Frege, *Grundgesetze der Arithmetik*, Bd. I, 1893, Bd. II, 1903. Frege's *Grundlagen* is a sort of commentary on the *Grundgesetze* and frequent references are found to the *Grundlagen*. For the student of the history of logic, these two volumes are extremely important.

point of view in affirming or denying that adding 2 to 3 gives 5, but merely that $2+3=5$ denotes some truth value. Thus $(\xi^2=4)=(3.\xi^2=12)$ denotes that $\xi^2=4$ has the same truth value as $3.\xi^2=12$. To denote that we have an assertion of a truth value, Frege introduces the symbol $\vdash$. (We shall meet this in other places as well.) Thus $\vdash 2^2=4$ means that the truth value denoted by $2^2=4$ is asserted: i.e., 4 actually is the square of 2. A judgment recognizes the truth of a thought, and the symbol $\vdash$ denotes that we are dealing with the symbolic representation of a judgment which is called a proposition (p. 9).

The symbol $—$ (not to be confused with a minus sign) when attached to a symbol Δ denotes that $—\Delta$ is truth whenever Δ is, and is falsity if Δ is not true. $—\Delta$ may be read "the truth-value of Δ." That is, if Δ denotes a true state of affairs (as, e.g., $2^2=4$), then $—\Delta$ denotes that we are dealing with the *proposition* $2^2=4$ and that the truth value of this proposition is truth; if Δ denotes a false state of affairs as, e.g., $2^2=5$, then $—\Delta=—2^2=5$ denotes that the truth value of the *proposition* $2^2=5$ is falsity. If Δ is a truth value then $\Delta=—\Delta$.

The value of $\top\xi$ is falsity if $—\xi$ is truth and truth if $—\xi$ is falsity. It follows that $\top\Delta=\top(—\Delta)=—(\top\Delta)=—\top(—\Delta)$. This is obviously the symbol for contradiction.

Thus $\vdash$ means we assert the truth of the proposition while $\vdash\!\top$ means we assert the denial of the truth of the proposition. For example,

$$\vdash 2^2=4$$
$$\vdash\!\top 2^2=5$$

In order to express universal propositions, i.e., propositions which state that a function is true for all values of the variable, Frege uses the symbol $—\smile—$; so that $—\overset{\mathfrak{a}}{\smile}—f(\mathfrak{a})$ means that $f(\mathfrak{a})$ is true for all values of the argument a. If that is actually the case then the truth value of $—\overset{\mathfrak{a}}{\smile}—f(a)$ is truth, if there is one value for which it is not the case, then the truth value of $—\overset{\mathfrak{a}}{\smile}—f(a)$ is falsity. Certain combinations are now possible.

$—\overset{\mathfrak{a}}{\smile}—f(a)$ means "for all a, $f(a)$ is true."

$\top\overset{\mathfrak{a}}{\smile}—f(a)$ means "it is false that $f(a)$ is true for all a." Thus $—\overset{\mathfrak{a}}{\smile}—f(a)$ and $\top\overset{\mathfrak{a}}{\smile}—f(a)$ are contradictions, i.e., if one is true the other is false; whereas $—\overset{\mathfrak{a}}{\smile}—f(a)$ and $—\overset{\mathfrak{a}}{\smile}\top f(a)$ are contraries, while $—\overset{\mathfrak{a}}{\smile}\top f(a)$ implies $\top\overset{\mathfrak{a}}{\smile}—f(a)$, i.e., these are subalterns.

$\top\overset{\mathfrak{a}}{\smile}\top f(a)$ means "it is false that $f(a)$ is false for all a," i.e., there exists at least one a such that $f(a)$ is true. The assertion symbol $\vdash$ can be attached to each of these as $\vdash\overset{\mathfrak{a}}{\smile}—f(a)$; $\vdash\!\top\overset{\mathfrak{a}}{\smile}—f(a)$; $\vdash\!\top\overset{\mathfrak{a}}{\smile}\top f(a)$; $\vdash\overset{\mathfrak{a}}{\smile}\top f(a)$.

If $-\!\overset{\mathfrak{a}}{\smile}\!-\phi(a)=\psi(a)$ is true, we can also say that the $\phi(\xi)$ has the same range of values as $\psi(\xi)$, i.e., we can replace a universal propositional equality by an equality of range of values. Suppose we have the two propositional functions intended as universals:

(1) "x is a mortal" and

(2) "x is a living thing."

These may be written as $-\!\overset{\mathfrak{a}}{\smile}\!-f\mathfrak{a}$ and $-\!\overset{\mathfrak{a}}{\smile}\!-g\mathfrak{a}$. The equality

(3) $-\!\overset{\mathfrak{a}}{\smile}\!-f\mathfrak{a}=-\!\overset{\mathfrak{a}}{\smile}\!-g\mathfrak{a}$

means that the truth value of propositional function (1) is the same as that of propositional function (2). Since each proposition is satisfied by the same range of values, we can replace equality (3) by a statement about their ranges of values. We denote the range of values of $\phi(\xi)$ by $'\epsilon\phi(\epsilon)$.

For example, if $\phi(\xi)=\zeta$ is a mortal, then $'\epsilon\phi(\epsilon)$ denotes *all men* taken in extension. It follows that $-\!\overset{\mathfrak{a}}{\smile}\!-\phi(a)=\psi(a)\equiv{}'\epsilon\phi(\epsilon)={}'a\psi(a)$. (The introduction of this is necessitated by the needs of the definition of number.)

We now introduce the symbol

$$\begin{array}{l}\top\!\!-\xi\\ \llcorner\!-\zeta\end{array}$$

which has falsity as its value whenever ζ is true and ξ is false, and truth in all other cases. It may be read (in those cases where ξ and ζ are propositions) "either ξ is true or ζ is false" or "if ζ is true ξ is true." (This is the definition of implication which Russell adopts.)

Thus we have $\begin{array}{l}\vdash\!\top\!-3^2>2\\ \quad\llcorner\!-3>2\end{array}$ which reads if $3>2$, then $3^2>2$, or since $\zeta\ (=3>2)$ is true and $\xi\ (=3^2>2)$ is also, the value of the subsumption is truth.

$\begin{array}{l}\vdash\!\top\!-1^2>2\\ \quad\llcorner\!-1>2\end{array}$ has truth for its value also since $1>2$ is false and $1^2>2$ is false. But $\begin{array}{l}\vdash\!\top\!-1>2\\ \quad\llcorner\!-3>2\end{array}$ has falsity as its valuc since $3>2$ is true and $1>2$ is false.

Combinations with the contradiction stroke are possible:

$\begin{array}{l}\vdash\!\!\!\!\!\top\!\!\top\!-\xi\\ \qquad\llcorner\!-\zeta\end{array}$ which reads "ξ is false *and* ζ is true." For example,

$\begin{array}{l}\vdash\!\!\!\!\!\top\!\!\top\!-2>3\\ \qquad\llcorner\!-2+3=5\end{array}$ which reads "$2>3$ is false *and* $2+3$ is 5."

$\begin{array}{l}\vdash\!\!\!\!\!\top\!\!\top\!\!\top\!-\xi\\ \qquad\llcorner\!\!-\!\!-\zeta\end{array}$ which reads "ξ is true *and* ζ is truc." For example,

$\vdash\!\top\!\!\top$ $3>2$ / $2+3=5$ which reads "3 is greater than 2 *and* $2+3$ is 5.

$\vdash\!\top\!\!\top$ ξ / $\top\zeta$ which reads "ξ is false *and* ζ is false," or "neither ξ is true nor is ζ true." The example $\vdash\!\top\!\!\top$ $2^3=3^2$ / $\top\, 1^2=2^1$ reads "neither is $2^3=3^2$ nor is $1^2=2^1$."

Since $\vdash\!\top\!\!\top$ ξ / $\top\zeta$ means ξ is false *and* ζ is false, if we were to add another negation stroke, i.e., the $\top$ to these, it would deny that the two are false and we would have that one at least is true. Thus $\vdash\!\top\!\top\!\!\top$ ξ / $\top\zeta$ reads either ξ is true or ζ is true, i.e., it is false, that ξ is false *and* ζ is false.

For example, $\vdash\!\top\!\top\!\!\top$ $3^2>3$ / $\top\, 3<3$ means either $3^2>3$ or else $3<3$.

We might write down the following equations:

$$\vdash\!\top\!\!\top\ \xi\,/\,\top\zeta \equiv \vdash\!\top\!\top\!\top\!\!\top\ \xi\,/\,\top\zeta;$$

$$\vdash\!\top\!\top\!\!\top\!\top\ \xi\,/\,\top\zeta \equiv \vdash\!\top\!\!\top\ \xi\,/\,\top\zeta \text{ etc.}$$

So far we have had only one lower and one upper term. It is possible to have many subordinate terms, since ξ can stand for anything and therefore for S / T —thus we could have $\vdash$ S / T / V which would read "if V is true then S / T is true," or expanded, "if V is true then if T is true S is also," i.e., the truth value of the complete symbol is *falsity* only when V is true and S / T is false, i.e., whenever V and T are true and S is false, then $\vdash$ S / T / V is false. Hence we can interchange the positions of V and T thus:

$$\vdash S\,/\,T\,/\,V \equiv \vdash S\,/\,V\,/\,T$$

It can also be shown that:

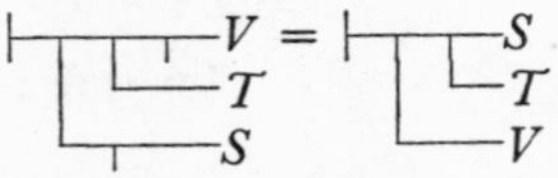

```
├─┬──┬┬─V = ├─┬─┬─S
  │  └──T     │ └─T
  └──┬──S     └───V
```

The negation stroke can be put into these expanded diagrams and read as before. A stroke preceding the letter denotes the denial of the letter. If it precedes the downward line it denies the subsumption which follows it. Thus:

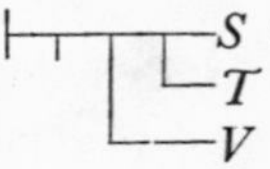

```
├┬┬──┬──S
     └──T
  └─────V
```

reads "V is true *and* T is true *and* S is false."

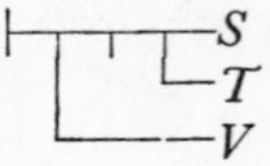

```
├───┬┬──S
     └──T
 └──────V
```

reads "either V is false *or* T is true and S is false.," etc.

We can now write the universal affirmative proposition as:

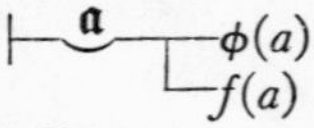

```
├─𝔞──┬─φ(a)
     └─f(a)
```

which reads "for all a, if $f(a)$ is true, then $\phi(a)$ is also."

For example,

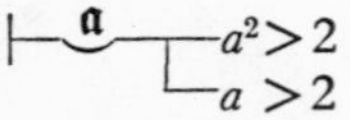

```
├─𝔞──┬─a² > 2
     └─a > 2
```

"for all a such that $a>2$, $a^2>2$ also."

Since we are interested in Frege's definition of number and not in his logic, we shall omit any discussion of his methods of drawing conclusions and continue to build up to his definition of number. We shall not go into detail here. It will be sufficient for our purposes to indicate how on the basis of the symbolism (logical) we have developed, it is possible to define number. We shall thus have an outline of the attempt to derive number from logic. This is followed in general by Russell-Whitehead also.

We introduce another definition here:

$$\vdash a = \backslash\text{'}\epsilon(a=\epsilon)$$

which denotes that the extension of the concept in question is one and only one object, i.e., a is *the* extension of the concept $\backslash\xi$.

$\Vdash$ denotes we are dealing with a definition.

We define:

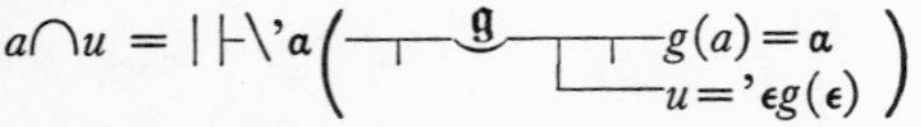

```
a⌒u = ||├\'α( ─┬──𝔤──┬┬─g(a) = α   )
                     └──u = 'εg(ε)
```

(This is the definition of an object being under the extension of a concept.)

That is, we are defining "$a \frown u$" as "the object coming under the extension of the concept $\grave{}\alpha\left(\overset{\mathfrak{g}}{\smile}\begin{array}{l} g(a)=\alpha \\ u={}'\epsilon g(\epsilon) \end{array}\right)$, this being the term

for an object α, so constituted that if u is the extent of the concept $g(\epsilon)$ then the value of this function for the argument a is equal to α irrespective of which function . . . g may denote,"[14] or in other words:
$a \frown u$ is the extension of the concept *there exists a u equal to the extension of g(a) such that for all g, g(a) is equal to* α."

$$Ip = \Vdash\left(\overset{\mathfrak{e}}{\smile}\overset{\mathfrak{d}}{\smile}\overset{\mathfrak{a}}{\smile}\begin{array}{l} d=a \\ e\frown(a\frown p) \\ e\frown(d\frown p) \end{array}\right)$$

(This defines 1,1 correspondence between two concepts.)
This denotes the univocalness of a relation. It reads in words, "The sign '*Ip*' is equivalent to the statement that a relation between the objects e and d, $[\phi(e, d)]$ and a relation between the objects e and a, $[\phi(e, a)]$ for any value of e, d, and a, results in the value for d being equal to the value for a."[15] In other words: *Ip* means that for all e, d, and a, the extension of the function corresponding to $e\frown(d\frown p)$ and the extension of the function corresponding to $e\frown(a\frown p)$ are so related that to each value d of one function one and only one value a of the other function corresponds.

Having defined a one-to-one correspondence we next define the process of setting up such a correspondence, i.e., the imaging (Abbildung) by means of a relation:

$$>p = \Vdash{}'\alpha{}'\epsilon\left[\overset{\mathfrak{d}}{\smile}\begin{array}{l} d\frown\epsilon \\ \overset{\mathfrak{a}}{\smile}\begin{array}{l} a\frown\alpha \\ d\frown(a\frown p) \end{array} \\ Ip \end{array}\right]$$

This denotes the establishment of a relation such that the objects comprised under one concept $d\frown\epsilon$ are coordinated *univocally* with objects $d\frown(a\frown p)$. In words, we can read this as "$>p$ is equivalent to those pairs of objects which are so constituted that any of the objects coming under the concept $d\frown\epsilon$ stands in a univocal relation to an object under the concept $a\frown\alpha$."[16]

In order to define number we must have one-to-one correspondence both ways. For this we need to define the converse of a relation: $\mathfrak{P}p = \Vdash{}'\alpha{}'\epsilon(\alpha\frown(\epsilon\frown p))$, which reads: $\mathfrak{P}p$ denotes the pairs of objects

[14] Jörgenson, Vol. I, p. 169.
[15] *Idem.*
[16] Jörgenson, *ibid.*, p. 170.

related to each other by $\alpha\frown(\epsilon\frown p)$ which is the converse $\epsilon\frown(\alpha\frown p)$.

We can define the quantity of a concept as:

$$Qu = \mid\vdash\text{'}\epsilon\left(\top\overset{\mathfrak{g}}{\smile}\top\begin{cases}\epsilon\frown(u\frown>g)\\ u\frown(\epsilon\frown>\mathfrak{P}g\end{cases}\right)$$

in words: Qu is equivalent to the objects which are reciprocally coordinated under the concepts $\epsilon\frown u$ and $u\frown\epsilon$.

Clearly the objects reciprocally correlated under the concept $\top x=x$ are null. Therefore the *quantity* 0 can be defined as:

$$\emptyset = \mid\vdash Q\text{'}\epsilon(\top\epsilon=\epsilon)$$

Likewise as explained above the number 1 can be defined as

$$I = \mid\vdash Q\text{'}\epsilon(\epsilon=\emptyset)$$

Having defined the general notion of number, and the numbers $\emptyset$ and I, all that remains is to define a method of generating a successor to any given number. This Frege defines as follows:

$$f = \mid\vdash\text{'}\alpha\text{'}\epsilon\left[\top\overset{\mathfrak{u}}{\smile}\overset{\mathfrak{a}}{\smile}\top\begin{cases}\top\begin{cases}Qu=\alpha\\ \alpha\frown u\end{cases}\\ Q\text{'}\epsilon\left(\top\begin{cases}\epsilon=\alpha\\ \epsilon\frown u\end{cases}\right)=\epsilon\end{cases}\right]$$

We might list the elements involved in this very lengthy development of the definition of number:

one-to-one correspondence
converse of a relation
extension of a concept
all *a*
negation
subsumption (or implication).

All of these are obviously logical concepts, defined in logical terms. The question which remains is whether number is adequately defined in these terms. This question we shall not discuss here.[17] In any case, even if this definition is ultimately accepted, since logic defines or states the properties of any objects whatsoever (i.e., the general structural properties of things), number would also express such properties.[18]

But one element is seen to be extremely important. The use of the "extension of a concept" introduces the notion of a *class*. Every class of objects is defined as the extension of a concept, i.e., as those elements to which one can apply the concept. This occurs for Frege in the symbol $\text{'}\epsilon\phi(\epsilon)$ and in the definition of $a\frown u$. This means that fundamental to

[17] *See* Jörgenson, *Treatise of Formal Logic*. Vol. 3, for discussion of this problem.
[18] This idea will be elaborated in the last chapter of this book.

mathematics lies the notion of a class. Thus it is necessary to examine the theory of classes (German: Menge) and build up from them to numbers. This was already done by Cantor, and has given rise to the *Mengenlehre* or Theory of Aggregates which we shall discuss briefly later.

Russell's indebtedness to Frege is great. The basic notions and method of attack of *P.M.* are found in Frege's work. This does not lessen the value of the contribution of Russell's for to discover a new symbolism that works is indeed a great achievement in itself.

CHAPTER FIVE

A. PASCH'S EMPIRICO-POSTULATIONAL DEFINITION OF NUMBER

The Peano-Russell definition has as its leading idea the derivation of the general concept of number, as well as the particular numbers, from pure logical concepts. Such a derivation is always on a level which seems not to consider the process of abstraction nor the concrete entities which later come to be the "stuffing" of the logical "crust." It takes its start from a high degree of abstraction without considering the process of arriving at this degree of abstraction.

It is possible to take another point of departure and consider mathematics and number as properties of objects, but a different kind of property from color, etc. Mill, however, had insisted that number was abstracted from objective groups and was a property of objects like their color.[1] We need not criticize this any more than to point out that Mill never could get to ordinary arithmetic on this basis. The leading idea of Mill, that number is ultimately derivable from what is experienced, appears again in the work of Moritz Pasch. Unfortunately, the work of Pasch, which we shall discuss in this section, was done at a time when the logistic investigations were proceeding at a rapid rate. Hence his work has been almost completely overlooked[2] (except for his work on geometry).

The works of Pasch important for our study are: *Grundlagen der Analysis*, Leipzig, 1909; *Mathematik und Logik*, Leipzig, 1919; *Der Ursprung des Zahlbegriffs* I, *Arch. Math. Phys.* Bd. 28 (1920) pp. 17–34, II *Math. Zeit*, Bd. 11 (1921). The two parts of *Der Ursprung des Zahlbegriffs* were published in one pamphlet, Berlin, 1930. In our discussion we shall follow this last cited work which I consider the most decisive work of Pasch.[3]

Pasch himself gives the names of "Empirismus" and "Kombinatorik" to his work. He feels that ultimately mathematics is a branch

[1] J. S. Mill, *Logic*, Book II, Chap. 6.

[2] A. Heyting, *Mathematische Grundlagenforschungen*, 1934, pp. 59–62, does summarize the leading ideas in Pasch's work. Cf. also G. Stammler, *Der Zahlbegriff seit Gauss*, 1926, pp. 117 ff.

[3] In Pasch's later work there is evident an attempt at greater precision. Also the term *Grundsatz* of the earlier work is replaced by *Kernsatz*. Cf. *Ursprung*, p. 1, note 1. KB stands for Kernbegriff, which means "root concept."

of the natural sciences based on what is experienced and tested by experience. Number itself can be constructed by combining certain concepts. These fundamental concepts are:

1. Entity (Ding)—experienced object
2. Event (Geschehnis)
3. Denotation of an entity (Angabe)
4. Proper name (Eigenname) (symbol)
5. Collective name (Sammelname)
6. Earlier or later event
7. Immediately succeeding event
8. Chain of events

From these concepts *number* can be derived. The nature of the process and of the fundamental concepts denotes the basis in experience of the number concept. What Pasch is attempting to do is to get at the very fountain from which the abstractive process takes its start. He wishes to go behind the logistic level to a more basic one.

Pasch feels that the structure of mathematics gives certainty only if it is derived from a root (Kern) which consists of a collection of root propositions (= axioms). The word *root* (Stamm; later Kern) is used to denote that the discussion concerns the substructure of a science. This substructure is described in the form of a collection of propositions.[4] Among the root propositions must be found everything used for proving any theorems.[5] The root of a system consists of root concepts and root propositions.[6] Such a construction for arithmetic will demonstrate the internal consistency of arithmetic (Haltbarkeit). Since arithmetic is the most fundamental branch of mathematics, it cannot be referred to any other branch but must be developed in itself. Since what we seek is the ground from which arithmetic springs, it is not sufficient to begin by using terms like "aggregate" (Menge) or "number" in their full present-day significance. The present-day meaning does not indicate the source of arithmetic,[7] which seems to be in some degree psychological. Arithmetic can be developed in any individual. "The development herein described is such that it can be developed by every human being if that person (1) considers only those entities which he himself experiences and observes to be distinct, and (2) when he allows himself unbounded life and unlimited thought."[8]

In other words, arithmetic seems to be based on the nature of experience and the ability to distinguish between entities.

---

[4] *Mathematik und Logik*, Leipzig, 1919, pp. 4–5, pp. 37–38.
[5] *Vorlesungen über neuere Geometrie*, 2nd ed. Berlin 1926, p. 4, also p. 248.
[6] *Ibid.*, p. 18.
[7] Cf. *Ursp. Zahl.* pp. 2–3.
[8] *Ibid.*, p. 6.

(Pasch's development has many similarities to Dedekind's work. The reader should consult Dedekind's essay on "Nature and Meaning of Number" in *Essays on Number*, London and Chicago, 1924.)

The root concepts have been listed above. I shall now explain them in more detail.

KB 1. *Entity.* This seems to be anything which occurs in experience or is capable of being experienced.[9]

2. *Proper Name.* A proper name is a name which denotes one and only one entity. The assignment of a proper name may be arbitrary.[9] A proper name seems to be the *symbol* for an entity.

3. *Denotation of an Entity* (Ein Ding angeben). By denoting an entity we mean any "pointing to" the entity.[10] This would appear to be the ostensive definition of the entity.

Pasch is trying to keep separate and distinct the object, the symbol which represents the object, and the essential nature of the object. In other words the distinction is between the symbol, the referent, and the definition of the symbol.

4. *Event.* An event is any denotation of an entity.[10]

5. *Collective Name.* A collective name is any name which denotes any specific entities, whose denotations have been given (e.g., *event* is a collective name).[10] The collective name corresponds to the class concept. It is not the same as the entities it denotes.

6. *Earlier, Later.* Since of two entities one must precede the other in experience, we call the one which precedes *earlier*—the other *later*.[11]

7. *Immediate Successor.* An immediate successor of a given event is recognized in the complete experience of the class of events to which it belongs.

8. *Chain of Events.* The chain $\mathfrak{A}$ of events A seems to be that concept which enables one to treat A as an entity. The emphasis should be on the *chain.* It appears to be the connecting concept, and not merely the collective name, of all elements. It arises only after we have the concepts of *earlier*, *later*, and *successor* and hence implies order. For example, two-legged animal may be considered the collective name of every individual human being. When we recognize the relation between men we may view the class *man* as a new entity. Then the concept *man* which relates individual men in order to constitute the ordered class

---

[9] *Urs. Zahl.* p. 6; also *Grundlagen d. Analysis*, p. 1.
[10] *Ursprung des Zahlbegriffs*, p. 7.
[11] *Ibid.*, p. 8.

is called the chain of the events *men*.[12] The *chain* seems to be that which makes of a collection an ordered class. The *chain* is to the *collection* what the *meaning* is to the *entity*.

Pasch has drawn a distinction here between the symbol for a class and the definition (intention) of the class. The extension of the class would be obtained by denoting the entities of the class. Thus one can denote the class as such and also the entities in the class.

Pasch says that these constitute all the root concepts necessary to construct arithmetic. All other concepts of arithmetic are definable or derivable from these.[13] We shall now list the fundamental definitions and root propositions which relate these concepts.

Definitions. (Df.) 1. The entity to which the proper name a is assigned is called the *meaning* (or *possessor*) of the name a. (Briefly; I shall speak of a.)[14]

2. Proper names assigned to the same entity are said to be *synonymous*.[15]

3. The denoted entity is said to be the *object* of the denotation.[15]

4. Those entities to which are assigned a collective name are called the meanings (carriers) of the name.[16]

5. Collective names assigned to the same entities are said to be *synonymous*.

6. If a is earlier than b, then we can also say: a occurs before b; a precedes b, or b happens after a; b follows a.

7. Let A be a collection of denoted events and let a be that meaning of A which occurs before any other meaning of A, a is then said to be the earliest A or the *first* A. Let b be the denoted element of A which occurs later than any other A, b is then said to be the *last* A.

8. If event a precedes event b and event c follows event b, then b is said to lie *between* a and c.

9. The A immediately following any e is also called: *the first A after e*; the successor in A to e.

10. By *event p* we mean: the last element A preceding q or the A immediate predecessor of q.

[12] *In Grund. d. An.*, *Folge* is used for *Kette* (p. 7). *Urs. Zahl.* p. 17.
[13] *Urs. Zahl.* p. 37, footnote 1.
[14] *Ibid.*, pp. 6 and 33.
[15] *Ibid.*, p. 34.
[16] All references are to pp. 33–41 in *Ursprung des Zahlbegriffs* and can be easily found.

11. The elements of A are called the *ingredients* (or proper parts) of the chain $\mathfrak{A}$ of A. I say; $\mathfrak{A}$ consists of events A.

12. If a and b are events of A, but no A lies between a and b, then a and b are said to be *neighbors* in chain $\mathfrak{A}$.

13. When I say that event g is later than $\mathfrak{A}$, I mean: g is later than all A.

14. If P is the collective name for an event a' and also for only one other event b', then the chain $\mathfrak{P}$ of events P is said to be a *dual chain* (Paarkette).

15. If events P fall under the collective name A having a chain $\mathfrak{A}$, then we call the chain $\mathfrak{P}$ of P a *dual chain of* $\mathfrak{A}$.

16. If the ingredients (links) of chain $\mathfrak{A}$ are denotations of completely different entities then $\mathfrak{A}$ will be called a *pack* (Rotte) of entities A or: the *pack* $\mathfrak{A}$ *of entities* A. The A are said to be terms of the pack $\mathfrak{A}$.

17. Instead of saying "the entity a is a term in $\mathfrak{A}$," I shall say: the pack $\mathfrak{A}$ contains the entity a.

18. The concepts "precedes," "first," "last," "between," "immediate successor," "immediate predecessor," "neighbor," can be defined for the terms of the pack $\mathfrak{A}$.

19. The first term is also called *initial term*, the last is called *end term*. Initial and end terms are said to be *external* terms, all others *internal*.

20. If $\mathfrak{P}$ is a dual chain, denotations a' and b', in which α' is the object of a' and β' that of b', then $\mathfrak{P}$ is also called a *dual pack*. If a' and b' are neighbors in $\mathfrak{A}$ then $\mathfrak{P}$ is a *neighbored pack* of $\mathfrak{A}$.

21. Any dual pack whose terms are terms of the dual pack $\mathfrak{B}$ and which begins with the same initial term with which B begins is said to be *conformal* to $\mathfrak{B}$.[17]

22. A dual pack which begins with the end term of another dual pack and ends with the initial term of the other is said to be the *converse* of the other.

23. To denote entity β after denoting entity α is to *pass* from α to β. To denote a second time the first term of a pack and then to pass from each redenoted term which is not the last to the immediate successor is to *enumerate* the pack.[18]

[17] Cf. Also *Grund. An.*, pp. 10–11.

[18] The word is "abschreiten." Pasch is obviously leading up to mathematical induction. The word "enumerate" is to be taken in the sense of "telling off" and should not be taken in the sense of counting.

24. If every neighbored pack of a pack $\mathfrak{S}$ is conformal to a neighbored pack of pack $\mathfrak{A}$, then $\mathfrak{S}$ is said to be conformal to $\mathfrak{A}$.

25. If we can say of an assertion k_ρ about the entity ρ, that if the assertion is valid for any term which is not the last of the pack $\mathfrak{A}$, that it passes from this term to the immediate successor, then k_ρ is said to be *an assertion enumerable in pack* $\mathfrak{A}$.

26. If the validity of an assertion is demonstrated for all the terms of the pack by means of Definition 25, then this proof is said to be *proof by enumeration.*

27. Of the entities A contained in pack $\mathfrak{A}$ we can also say that "they are contained in entity M" under the condition that we do not speak of M in any other connection. M is a *collection.*[19]

28. If α and β are the only entities subsumed under the collective name A, then M is said to be a *pair.*

29. M of Definition 27 is called "the" collection of entities A. The entities are called the *parts* of the collection. (At this point a theorem can be proved "the collection is not a part of itself" which gets rid of one of the paradoxes to be noted later. But the question remains as to whether one can define away a difficulty.)

30. By "the terms of the pack $\mathfrak{A}$ from a to b" we mean a, b, and all terms lying between a and b.

31. If B is the collective name for the terms of the pack $\mathfrak{A}$ from a to b, a preceding b, then the chain of denotations of entities B in $\mathfrak{A}$ is called a *segment* of $\mathfrak{A}$, i.e., the segment from a to b.[20]

32. If we can assert of a proposition concerning a segment of the pack $\mathfrak{A}$ which begins with the first term that if the proposition holds for any segment, it is transferable to the segment which reaches to the immediately following term, then this proposition is said to be "a proposition about the segment enumerable in the pack $\mathfrak{A}$."[21]

---

[19] "The concept of a collection is first reached after things are individually denoted, i.e., after we have a pack of entities. No doubt, it happens that anything first appears to me to be a whole and is later analyzed into parts. But only after I have recognized these individual things—here a, b, c—do I gain a distinct concept of the entity G, which is the collection of the things a, b, c.

"The entity G (collection of entities a, b, c) is different from the entities a, b, c." (*Urs. Zahl.* p. 28.) This means that a collection is not its elements.

The definition of collection here given is an implicit definition. Cf. p. 40, *Urs. Zahl.*

[20] Segment = Abschnitt. As here used it appears to be a subclass. Cf. *Urs. Zahl.*, pp. 40, 31, and *Grund An.*, pp. 8–10.

[21] Pasch clearly means proof by mathematical induction. Cf. p. 27.

33. The process of redenoting the last term and transferring from each redenotation which is not the first to its immediate predecessor is said to be *converse enumeration*.

34. If 𝔅 is the segment of pack 𝔄 which contains all terms of 𝔄 except z, then we can say instead of "constructing a pack 𝔄" — *add* z *to* 𝔅. If we can assert of a proposition about a pack that the proposition is valid for the pack which results when a term is added to the given pack, then we call the proposition—an *enumerable proposition concerning packs*.

35. To *add* z *to the collection* M means to construct the collection which contains all parts of M and in addition only the entity z.

36. If 𝔄 and 𝔅 are packs of the same entities and if the neighbored packs of 𝔅 are, respectively, *converses* (see Definition 22) of the neighbored packs of 𝔄, then 𝔅 is said to be the converse of 𝔄.

37. If a proposition about an entity is such that when it holds for any term, not the first, of the pack 𝔄 it transfers to the immediate predecessor, then that proposition is said to be *inversely enumerable in* 𝔄.

These definitions are all needed by Pasch to lead up to the definition of aggregates and then of number. In addition there are needed a number of root propositions which I now give. From these we shall proceed to our definition of the natural numbers.

K.S.[22] 1. I can assign to any entity a proper name. I cannot assign to another entity the same proper name. The proper name of an entity is itself an entity different from the original entity.[23] (K.S. 1 shows the essential dependence upon the I for the construction of mathematics. In *Grundlagen d. Analysis*[24] this element is further emphasized when Pasch says "Only what is perceived (Wahrgenommenes) or what is perceivable can be an entity.")

2. Many proper names can be assigned to the same entity. (Cf. Definition 2.) (The reader should try to prove: If a and b are synonomous and b and c are also, then a and c are synonomous.)

3a. I can denote any entity.

3b. The denotation may occur by means of a proper name assigned to the entity.

3c. Two entities cannot have the same denotations.

---

[22] K. S. is short for "Kernsätze"—which Pasch uses in place of axiom.

[23] This is very like the way Dedekind establishes the existence of an infinite class. But see remark after K. S. 7.

[24] P. 1.

3d. The denotation of an entity is itself an entity different from the entity it denotes as well as from its proper name.

3e. The denotation is an event.
(K.S. 3d says that we have an entity, its symbol, and its definition and these are all different. For example. If *A* and *B* are symbols for points, then the points, the letters *A*, *B*, and the definitions of *A* and *B* are all different, although they are all entities. Or the entity 5, the symbol 5, and the definition of 5 will all be different. It might be added that logisticism, like formalism, confuses these three.)

4a. Two or more denotations can be given for the same entity.

4b. I can assign a collective name to entities which have been denoted. This collective name is an entity different from the other entities. (4a means that an object can be defined in more than one way, but by 3c there can be only one object for a given definition. 4b implies that a collective name cannot be an entity subsumed under itself.)

5. Two or more collective names can be introduced for the same entities. (In the usual logical terms, this means that two classes may be identical in extension yet differ in intention.)

Beginning with Definition 6 and with K.S. 6, we have what Pasch calls *arithmetical concepts*, since these lead to the concept of number. The relations between these concepts form the object of arithmetic (theory of numbers).[25] It is to be noted that most, if not all of the concepts used above are ultimately logical in character. But in addition, the actual recognition of entities and the actual denotation of the entity is done by an "I." This ability to denote entities which the "I" possesses appears to be extremely important for Pasch. This is further evident since the concepts "earlier," "later," which form the basis of *order* are derived from the fact that we cannot have simultaneous events.[26] Pasch appears to retain the Intuitionist emphasis upon the activity of the mind, the Empiricist emphasis upon the abstract character of number, and the Formalist emphasis upon the combinatory character of mathematics.

6. If *a* is an event and *b* is another event, then either *a* is earlier than *b* or *b* is earlier than *a*.

If *a* is earlier than *b*, *b* is not earlier (=later) than *a*.

7. I can denote an entity not yet denoted after every event.

Let *C* be a collective name for any entities whatsoever which are

[25] Cf. *Grund. An.*, p. 2.
[26] *Urs. Zahl.*, p. 8.

assumed to be denoted. I assign to these denotations the collective name *A*. Denotations *A* of entities *C* are experiences and *I have observed* that it is possible for me to experience events after any events; therefore I assert that this is possible for every future event.

As Pasch emphasizes, observation and experience are fundamental in the very act of denotation and the recognition of order. It is ultimately through observation and experience that we are enabled to note what element comes first in a collection.[27]

The reader should try to demonstrate the theorems:

1) If events *A* among which are *a*, *b*, *c*, are denoted and if *b* lies between *a* and *c*, then *b* is neither the first nor the last *A*.

2) There exists only one entity *A* which is the first (last) entity.

3) The first and last entities are different.

8. If *e* is any of the events *A*, not the last, then I can denote one and only one *A* immediately following *e* among the *A*.

If *g* follows *e* in *A* and *h* follows *g*, then *h* is not the immediate successor of *e*.

By K.S. 7, the activity of denoting is limited only by the span of life of the denoter, i.e., it may be said to have no intrinsic limit. But any collection of events *A* is always finite since every entity in a collection has been denoted. Furthermore, by K.S. 8, the events *A* form a discrete series, i.e., each term has one immediate successor. These properties are all properties of the natural number series. But in *Grundlagen d. Analysis*, page 79, Pasch defines the number *infinity* as follows: "If there are arbitrarily many entities having class name *D* (i.e., we can denote more than *n* entities *D* no matter how great *n* is), then we say there are *infinitely* many *D*."

This does not contradict what was said since the act of denoting can always go one step further. This infinite merely means the *ability* to go one step further. It is not an actual infinite since the *denoter* is finite. But the definition of infinite collections in *Grundlagen d. Analysis*, page 94, does seem to involve contradictions. (In *Ursprung des Zahlbegriffs*, no mention is made of infinite in any sense, except that on page 43 the remark is made that given any collection there exists a collection greater than it.)

9. If events *A* are denoted and *q* is any *A* not the first, then I can denote an *A* such that *q* is its immediate successor.

10. If a denotation *a* of an entity **a** has occurred, then there can also occur a new denotation of the entity **a** after the denotation *a*.

If *a* and *b* are denotations of different entities such that *a* is earlier than *b*, then we can denote the object *a* again with a denotation later than that of *b*.

[27] *Ibid.*, p. 9, also p. 13, paragraph 3 of §7.

The reader should now try to prove:

Entities can be denoted after denotations A of entities C such that no one of these are subsumed under the collective name C.

11. If events A are denoted, then one and only one entity $\mathfrak{A}$ can be denoted which is the chain of A.

$\mathfrak{A}$ is not the chain of other events also. (Determine when two chains are identical.)

12. The chain is no event.

13. If $\mathfrak{A}$ is any pack, then it is possible to enumerate it after $\mathfrak{A}$. This is also possible after an event e which follows $\mathfrak{A}$.

This enumeration yields all the terms of $\mathfrak{A}$ as the terms of a new pack $\mathfrak{A}'$.

The neighbored packs of $\mathfrak{A}'$ are conformal to those of $\mathfrak{A}$.

The root of arithmetic is now completed. Actually, it comprises definitions 1–23 and these K.S. Definitions 24–37 can be constructed from Definitions 1–23.

Certain theorems are important and should be derived by the reader:

1) If proposition K_r is valid for the first term of a pack $\mathfrak{A}$ and the proposition is enumerable in $\mathfrak{A}$, then it is valid for every term of the pack.

2) A collection is not part of itself.

All other concepts are derived from these roots and give us the entire structure of arithmetic.[28] We now proceed to derive the number concept.

If we have entities A and B denoted then we can derive pairs H such that (1) every pair is composed of one A and one B; (2) no A or B is repeated in the pairs; (3) if I desire to construct another pair from A and B, then I shall have to repeat either an A or a B or both.

This process of constructing these pairs is called *correlating*.

If in H all A and all B occur, then A and B are said to be *totally correlated*.

If all A occur in H but not all B, then B is said to be *excessively correlated* to A and A *defectively correlated* to B.

If the correlation between A and B is total and A is different from B, then A and B are said to be *equipotent*. If B is excessively correlated to A, then B is said to be *stronger* than A and A is *weaker* than B.

Let any collection whatsoever be denoted and let us assign to them the collective name M. Let $\mathfrak{Z}$ be a pack stronger than an M. Let $\mathfrak{Z}'$ be

---

[28] The following is taken from *Ursp. Zahl.* (pp. 42–44), but is developed in detail in *Grund. d. Analysis*, pp. 13–24.

the collective name of the terms of $\mathfrak{Z}$, and let *e* be the proper name of the first of these. If $\mathcal{N}$ is any M or, for that matter, any collection no stronger than $\mathfrak{Z}$, then I can denote one and only one entity, say *n*, following *e* such that the segment of $\mathfrak{Z}'$ up to *n* is equipotent with the collection $\mathcal{N}$.

The entity *n* is called the *number* derived from the pack $\mathfrak{Z}$ for the collection $\mathcal{N}$. I can use every $\mathfrak{Z}$ except *e* as a *number*. And I can use any M as the number collection.

The names one, two, three, etc., and the symbols 1, 2, 3, etc., are assigned to the numbers in succession.

The reader will see how close this definition of number is to the logistic definition. It is a valuable exercise to attempt to reduce the concepts and processes used by Pasch to those of Russell. The most fundamental difference is the recognition on the part of Pasch of the part the activity of the mind plays in arithmetic constructions.

One thing, Pasch feels, this construction does which no other does, i.e., it makes unnecessary the question of the consistency of arithmetic. The very nature of the construction makes contradiction impossible, in his opinion. Hence it is possible to demonstrate consistency by arithmetical interpretations and to explain the relation of mathematics and experience.[29]

B. THE PURE POSTULATIONAL DEFINITION OF NUMBER

The definitions of number given so far sought to derive number (1) from logical concepts; (2) from entities through empiristic abstraction. Each sought to find the ultimate roots of number (natural).

Meanwhile there was before each investigator the entire field of the results of arithmetic and the example from geometry of Euclid's Elements, i.e., the form of a set of postulates, axioms, and theorems. In 1899 David Hilbert published the first edition of his *Foundations of Geometry*, which gave a definitive postulational form to Euclidean Geometry including the assumptions which Euclid himself had omitted but used. It was inevitable therefore that this same technique be applied to arithmetic. The problem was to find a set of postulates which would enable one to *deduce* all the facts of arithmetic. The method consists of:

1. The enumeration of *all* fundamental *concepts* and fundamental *relations* of the science to be axiomatised. All additional concepts are to be defined in terms of these.

---

[29] *Mathematik und Logik*, *op. cit.*, pp. 17–18.

2. The statement of *all* axioms, i.e., propositions accepted without proof. All additional propositions are to be deduced from these by strictly logical methods.[30]

Clearly such a method presupposes that we have a great many results already before us which we desire to systematize. Such, of course, was the case with arithmetic. Further, such a definition takes for granted the entity to be defined and seeks merely to give its most important properties. It is not interested in its roots or origin. Such a definition therefore would not be incompatible either with the logistic or Pasch's empiristic derivation of number. Rather it would begin at the place where both of these methods have defined number in terms of some preceding concepts. However, the method does lead to a new point of view. Since in the postulates only properties of the entity are given, the entity is conceived as something about which nothing is known except that it satisfies the postulates stated. Hence the entity is introduced as a pure symbol which is *implicitly defined* by the postulates. Thus a number is merely a symbol which has certain properties given it by the postulates. It is then manipulated in accordance with the postulates. This gives rise to the Formalistic Foundations of Mathematics, which we shall discuss in more detail later in connection with David Hilbert and the Axiomatic Method.[31]

Probably the first axiomatic definition of number was given by Peano, in 1891. There was a good deal of discussion at the start of this century on the axiomatic definitions of number. (The reader should consult the *Bibliothèque du congrès international de philosophie*, Paris, 1900, Vol. 3.) Since then many definitions (axiomatic) have been given of the various types of numbers—natural, rational, irrational, transfinite, etc. (these terms will be made clear later). The best known of these to American readers are the postulate sets of E. V. Huntington, in the *Transactions of the American Mathematical Society*, Volumes 3 and 4, 1902, and 1903.[32] Huntington also gives citations to the early literature in the footnotes to these articles. (The reader should consult the bibliography of Fraenkel's "Einleitung in die Mengenlehre" for other references, as well as various issues of the *Bulletin of the American Mathematical Society*.) A set of postulates was given by David Hilbert in his article "Uber den Zahlbegriff" in *Jahresbericht der Deutschen Mathematiker-Vereinigung*, Volume 8, 1900.

---

[30] Heyting, *Math. Grund.* p. 30. Cf. also Jörgenson, *Treatise of Formal Logic*, Vol. 3, pp. 141 ff. Stammler, *Zahlbegriff seit Gauss*, pp. 124 ff. Fraenkel, *Einleit, in die Meng.*; F. Enriques, *Historic Development of Logic*, p. 174, and L. O. Kattsoff, "Postulational Techniques, III," *Philosophy of Science*, Vol. 3, July, 1936, pp. 375–417.

[31] Cf. Hilbert and Bernays, *Grundlagen der Mathematik*, Berlin, 1934, Vols. 1 and 2.

[32] The reader will do well to consult Huntington's admirable little volume, *The Continuum*, Cambridge, 1929, for examples of postulational definitions of various types and series.

Before beginning we must draw a distinction between a postulate set which says "given the natural number series 0, 1, 2, . . . , the properties it has are the following," and one which says, "any entity which satisfies the following properties will be called a natural number and will be denoted by one of the series of symbols 0, 1, 2. . . ." The first is the attitude adopted by Brouwer and the *Intuitionists*, the second by Hilbert and the *Formalists*.[33] Although the Formalists have used the postulational method almost without exception, the use of postulational technique is not a fundamental point of difference between the two schools.

First we present an axiomatic definition of natural numbers based on that of Peano given by Ludwig Neder in the *Jahresbericht der Deutschen Mathematiker-Vereinigung*, Volume 40, 1931, pages 22 and following, in an article entitled "Uber den Aufbau der Arithmetik."

Given a system of entities called natural numbers and denoted by small Latin letters. Between these entities there hold two fundamental relations: Equality—denoted by $a=b$ (read "a equals b"); Sequence (immediate successor)—denoted by $a<b$ (read "a precedes b"). (These relations are further defined in the postulates.)

There are to be four groups of axioms:

1. Axioms of equality and sequence (denoted by *G.F*)
2. Axioms concerning the Null (denoted by *N*)
3. Axioms concerning the existence of a successor (denoted by *E.F*)
4. Axioms concerning complete induction (denoted by *I*)

(The reader should note the recurrence in every definition, of the idea of induction. As a matter of fact, the natural numbers have been called the *inductive* numbers. Cf. Russell, *Introduction to Mathematical Philosophy*, p. 27.)

AXIOM SET

G.F. 1. $\left\{\begin{array}{l} a_1 < b \\ a_2 < b \end{array}\right\}$ gives $a_1 = a_2$

(read "if b is the immediate successor of both a_1 and a_2, then $a_1 = a_2$")

G.F. 2. $\left\{\begin{array}{l} a < b_1 \\ a < b_2 \end{array}\right\}$ gives $b_1 = b_2$

G.F. 3. $a = b$, $b < c$ gives $a < c$

G.F. 4. $a < b$, $b = c$ gives $a < c$

[33] Cf. Kattsoff, "Postulational Techniques, I," *Philosophy of Science* (Vol. 2, April 1935, pp. 134 ff.)

N. 1. There exists a natural number 0 such that if

$$a < b, \text{ then } b \neq 0.$$

E.F. 1. For every natural number a there exists at least one natural number b such that $a < b$.

I. 1. If a class K of natural numbers has the property:

1. 0 belongs to K
2. for every $a < a'$ for which a belongs to K, a' does also; then K contains every natural number.

(Certain properties of postulate sets will be explained later. Also, difficulties involved in the use of a postulate set itself will be discussed.)

H set for Positive Integral Numbers (*Trans.*, Vol. 3, p. 280). Huntington offers us alternative sets—the implications of this will be discussed when we talk about postulational techniques. These alternative sets are complete, i.e., they are (1) consistent = lead to no contradictions; (2) there is only one *such* group of postulates possible; (3) no postulate is a consequence of the others.[34]

H. 1. *Fundamental Concepts*:[35] A *class* of objects is determined when any condition is given such that every object in the universe must either satisfy or not satisfy the condition.

Every entity which belongs to the class is called an *element* of the class.

A *rule of combination* in a class is any rule or agreement by which, when any two elements are given in a definite order, some object is uniquely determined. If a and b are given, then the new object is $a \circ b$ (read "a with b").

$x = y$ means x and y denote the same object.

$x \neq y$ means x and y do not denote the same object.

POSTULATES OF MAGNITUDE[36]

1. If a and b are any elements of the class, then $a \circ b$ is also.

2. $a \circ b \neq a$

3. $(a \circ b) \circ c = a \circ (b \circ c)$ whenever a,b,c are elements of the class.

4. Whenever $a \neq b$ then at least one of the following conditions is satisfied:

1°) there is an element x such that $a = b \circ x$

2°) there is an element y such that $a \circ y = b$

[34] Huntington, *op. cit.*, Vol. 3, p. 264.
[35] *Ibid.*, pp. 266–67. Some changes are made here in the wording of the postulates.
[36] *Ibid.*, p. 280.

5. Let $a < b$ mean an element y exists such that $a \circ y = b$; let $a \leq b$ mean "$a < b$ or $a = b$."

If S is an infinite sequence of elements (a_k) such that

$a_k < a_{k+1}$, $a_k < c$ (where c is some fixed element)

then there is at least one element A having the following properties:

1°) $a_k \leq A$ when a_k belongs to S

2°) If y and A' are such that $y \circ A' = A$, then there is at least one element of S, say a_r, for which $A' < r$;

6. There is an element E such that $x \circ y \neq E$ whenever $y \neq E$. If o is interpreted as $+$ the above is a set for integral positive numbers. (It would be of value to the reader to examine the mathematical journals for other definitions of number in terms of postulate sets and compare them. The reader should also try to deduce consequences from the above two sets, and to see if the N and H sets are equivalent.)

CHAPTER SIX

EXTENSION OF THE NUMBER SYSTEM

We are not concerned here with the historical roots of the various types of number.[1] Our interest lies in finding and *inserting* into our number system all possible constructions of new types of number we can find. So far we have the series of what are called *natural numbers*—0, 1, 2, 3, 4,.... It has been said that God created the natural number series and man created all other numbers. We wish to retrace the process of creation. In doing so, it will also be necessary for us to assume that the fundamental arithmetical operations have been defined.[2]

Two trains of thought can demonstrate the necessity for new types of numbers: (1) algebraic, (2) geometric. The algebraic method necessitates the introduction of new types of numbers in order to make all its results meaningful. The geometric method shows the necessity of new types of numbers in order to make measurement meaningful for all cases. Of course, since the discovery of analytic geometry, these two methods have been seen not to be distinct. The algebraic method rests on the assumption which has been called "The Principle of No Exception" (also called the Principle of Permanence) which is stated as follows: "In the construction of Arithmetic, every combination of two previously defined numbers by a sign for a previously defined operation shall be invested with meaning, even where the original definition of the operation used excludes such a combination; and the meaning imported is to be such that . . . the old laws shall still hold good and may still be applied to it."[3] This really amounts to saying that the number system must be so extended that the result of any operation on two or more numbers will result in another number, i.e., the number system is to be a *closed* system. It is easily seen, for example, that the natural number series is not closed with respect to division.

---

[1] The reader should consult T. Dantzig, *Number, the Language of Science*, N. Y., 1930; L. Brunschvicg, *Les Etapes de la Philosophie Mathématique*, Paris, 1929.

[2] Cf. B. Russell, *Principles of Mathematics*, pp. 117–21. Neugebauer's *Vorgriechische Mathematik* is a recent book on the very early history of mathematics. A good discussion on a more technical level of the ground covered in this chapter will be found in E. W. Hobson, *Theory of Functions of a Real Variable*, 2nd edition, Cambridge, 1921, Vol. 1, pp. 1–58. The student would do well to work through this chapter in Hobson after reading our discussion. For a more general approach to these ideas the excellent book of R. Courant and H. Robbins, *What Is Mathematics* (Oxford Univ. Press, 1941) is worth studying.

[3] H. C. S. Schubert, *Mathematical Essays and Recreations*, Chicago, 1898, p. 14, and Dantzig, *op. cit.*, p. 92.

The geometric method rests on the assumption that there should correspond to every point on the line some number. This does not hold true with respect to imaginary and transfinite numbers—although we can give a geometric representation of imaginary numbers.

It must be kept in mind that the method of discovery does not carry with itself, necessarily, the logical construction of the new system. It is necessary, having discovered the possibility of a new type of number in one way or another, to reconstruct the logical base of the number system in order to have a deductive science of number, i.e., a science in which the new number is seen to be consistent with the old number and the fundamental postulates of arithmetic, and in which the new number is obtained from the old either by specialization or by generalization. (Cf. the appendix to this chapter.)

NEGATIVE NUMBERS[4]

1. *Algebraic Discovery*: Consider the linear equation $x + a = b$. Then we know that $x = b - a$. In this equation, a and b are taken to be natural numbers. Now so long as b is greater than a, i.e., occurs later in the series of natural numbers than does a, all is well. But suppose b is less than a: We have a situation which cannot be interpreted on the basis of the natural numbers alone. In order to give meaning to this equation $x = b - a$ (or, to be specific, say, $5 - 7$), we introduce a new type of number called the *negative* number. These negative numbers are denoted by placing before each natural number a line: $-1, -2, -3, -4, \ldots$

A great many pseudo-problems are avoided if it is recognized at once that the positive and negative numbers are only *opposites* and there is no question here of "negation" (Verneinung). As a matter of fact, Russell treats positive and negative numbers as relational numbers different from the cardinal number itself. If we recall that for Russell the numbers n and $n + 1$ are properties of classes, then there is a relation R such that $n\ R\ n+1$ and a converse relation $\check{R}$ such that $n+1\ \check{R}\ n$. The relational number of $n\ R\ n+1$ is $+1$, that of $n+1\ R\ n$ is -1. Of course, the rules of operation between the relational numbers must be defined. Since it can be demonstrated in the logical theory of relations that the converse of the converse of R is R, it follows from Russell's theory that the product of $(-n)$ by $(-n)$ is $+n^2$.[5] The disappearance

[4] Cf. Russell, *Introduction to Mathematical Philosophy*, pp. 63 ff. O. Hölder, *Die Mathematische Methode*, Berlin, 1924, pp. 194 ff., F. Waismann, *Einführung* in *das Mathematische Denken*, Vienna, 1936, pp. 3 ff., F. Kaufmann, *Das Unendliche* in *der Mathematik* and *Seine Ausschaltung*, Leipzig and Vienna, 1930, pp. 106 ff., and G. Stammler, *Der Zahlbegriff seit Gauss*, 1926, pp. 102 ff.

[5] The proof is as follows: A relation R is defined as the class of entities which are so related, i.e., $R = \hat{x}\hat{y}\ (xRy)$. The converse of the relation is $\check{R} = \hat{x}\hat{y}\ (yRx)$.

Let $\check{R} = R_1$ then $R_1 = \hat{x}\hat{y}\ (yRx)$ and $\check{R}_1 = \hat{x}\hat{y}\ (xRy)$, but $\hat{x}\hat{y}\ (xRy) = R$, therefore

$\check{R}_1 = R$ and since $R_1 = \check{R}$, it follows that $\check{\check{R}} = R$.

of the negative signs is the result of reducing negative numbers to relations and considering the product as taking a converse of a converse. Algebraic multiplication is a special case of logical multiplication. Such questions therefore as how one can add -3 to -3 a negative number of times result because of confusion between cardinal and relational numbers. Russell's definition of negative numbers is an attempt to set up the theory of negative numbers and is not the way negative numbers arose.

Once negatives are seen to be the opposite of positive numbers, then the negative is as much positive as the positives. That is, the distinction is a relative one, and, taken separately, the properties of both are the same. Hence we can transfer from the positive numbers such properties as $ab = ba$, to the negative numbers. However, as Hölder[6] points out, it is necessary to prove that the introduction of new numbers negative to those we have will not, on the basis of the laws of operation, lead to contradictions. Such a proof is not necessary, however, from Russell's point of view.

2. *Geometric Discovery*: Suppose we desire to correlate the natural numbers with points on a line. In order to do this, it is necessary to adopt a length which shall be our unit of measurement. Let -|——|- be our unit length. Then, laying this off along the line, we get

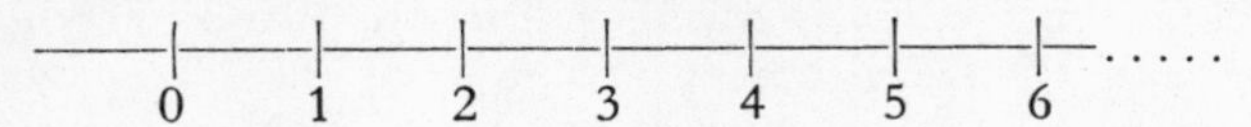

At the end of the first segment, we write 1, and then at the end of each segment thereafter the next number and so on. But this is a line segment and can be extended to the left as well, marking off segments. We denote segments in the opposite direction

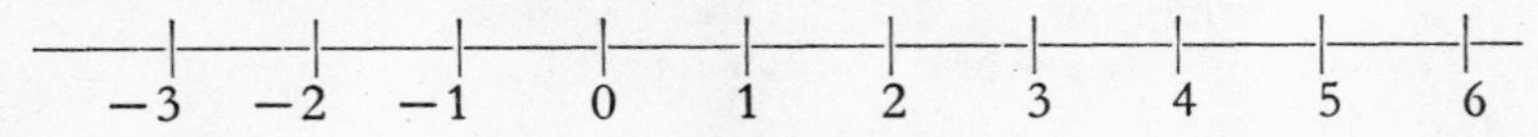

by $-1, -2, -3, \ldots$, for obvious reasons.

The properties of negative numbers are the same as those of natural numbers.

Numbers of the series of positive and negative numbers are called "*whole* numbers." The whole number series has no first or last number, and is closed with respect to addition and subtraction.

FRACTIONS

1. *Algebraic Discovery*: Given the equation $ax = b$, where a and b are natural numbers, then $x = b/a$. Whenever b is a multiple of a, the result is again a whole number. But suppose b and a are prime, i.e., have no factors in common, and b is less than a. We have then a

[6] Hölder, *op. cit.*, pp. 194 ff. attempts to prove this.

symbol composed of whole numbers but itself not a whole number. This symbol is taken to represent a new type of number called a *fractional* number. Fractions are not really new numbers but are relations between whole numbers. Russell actually defines the fraction m/n "as being that relation which holds between two inductive (= natural) numbers x and y, when $xn = ym$."[7] Fractions may therefore be treated as couples of natural numbers and their operations defined as follows:

(If $b = b' = 1$, we have the whole numbers.)
a) Equality: Two couples (a, b) and (a', b') are equal if $a.b' = a'.b$.
b) Addition: $(a,b)+(a'b') = [(ab'+a'b), bb']$.
c) Multiplication: $(ab).(a'b') = (aa', bb')$.

Thus fractions can be defined in terms of couples of *whole numbers*.

2. *Geometric Discovery*: The necessity of fractional numbers is seen from the fact that in correlating points to numbers as we did above, we note points between the whole number points. Thus if we divide the segment 0 to 1 into three equal parts, the dividing points can be denoted as 1/3 and 2/3. Since between any two points there is always another point and it can be shown that between any two fractions there is always another,[8] it would appear that all the points are exhausted. That this is not true will be evident below.[9]

Since between *any* two fractions, there are always other fractions, there is no fraction which is *next* to any given fraction. Any series which has the property that between any two of its terms there are always others, and that there is no *next* term to any term, is called a *compact* series. Thus the series of fractions is a *compact* series.

The series of numbers composed of the whole numbers and the fractions is called the series of *rational* numbers.[10] Again, here the word rational is not to be given a meaning it does not have. It means *ratio*, or *relation*. Perhaps we should call these *ratio-nal* numbers.

A rational number is any number which can be expressed as a ratio of two whole numbers p and q.

Rational numbers form a compact series, the whole numbers do not. Compact series are difficult to imagine and hence we begin at this point to leave the sphere of imagination and deal only with pure concepts. However, as we shall see later, it will be possible to rearrange the series of rational numbers, and correlate them with the natural numbers. There will then be one and only one rational number correlated with each natural number, and conversely one and only one

[7] *Introduction to Mathematical Philosophy*, p. 64. Of course when we treat numbers as pairs in this way, we are adopting a new point of view.

[8] Between the fraction a/b, and c/d, there is always the fraction $(ad + bc)/2(bd)$ lying midway between them.

[9] For the relations between fractions and measurements, *see* Kaufmann, *op. cit.*, pp. 110 ff.

[10] Waismann, *op. cit.*, p. 5.

natural number for each rational number. In this sense, there are as *many* rational numbers as natural numbers, although the properties of the two series are different.

IRRATIONAL NUMBERS

1. *Algebraic Discovery*: Although historically, irrational numbers seem to have been discovered first geometrically, we can continue our own development.

Suppose we have an equation $x^2 - a = 0$, where again a is a constant. Then $x = \sqrt{a}$. Obviously, if a is a perfect square, then x is a whole number. Suppose a is not a perfect square. Then x is not a whole number. Is it a rational number? Can it be expressed as a fraction p/q where p and q are whole numbers? The answer is already given in the tenth book of Euclid's geometry and was known to the Pythagoreans (a Greek sect whose fundamental doctrine was that number is the root of the universe). The proof for $a = 2$ is so simple it can easily be given.

We prove it by showing that if we assume $\sqrt{2}$ to be rational, an absurdity results. This is known in logic as a *reductio ad absurdum* proof and is based on the fact that every proposition is either true or false and there is no intermediate between truth and falsity. We shall argue that since the assumption that the square root of 2 is rational leads to absurdity, it must be false and therefore its opposite is true. Keep this argument in mind for our later discussion of the work of Brouwer.

Assume $\sqrt{2} = p/q$ (p and q relatively prime), then $2 = p^2/q^2$ and $p^2 = 2q^2$. This means that p^2 is even. If p^2 is even, p must also be even since the product of a number by itself can be even only if the number is even. If p is even, let $p = 2r$. Then from substitution, $4r^2 = 2q^2$, or $q^2 = 2r^2$, and q also is even. But we assumed p and q relatively prime, now they are seen not to be. A contradiction—and our assumption that $\sqrt{2} = p/q$ is false.[11]

Two things must be kept in mind:

a) Irrational here means merely non-ratio-nal. It has none of the connotations of the word "irrational" of ordinary discourse.

b) The fact that irrational numbers are seen to be necessitated does not mean that they have been properly introduced without contradictions.

2. *Geometric Discovery*: The geometric discovery was tied up with the right angled triangle and the famous Pythagorean theorem that the square on the hypotenuse was equal to the sum of the squares on

---

[11] For a proof of the general case see G. H. Hardy, *Course of Pure Mathematics*, 7th edition, 1938, p. 7. Kaufmann, *op. cit.*, arrives at these numbers through the process of extracting roots in general. pp. 120–21.

the other two sides. Thus if we have a right angled triangle in which the sides of the right angle were equal and both of length 1, :

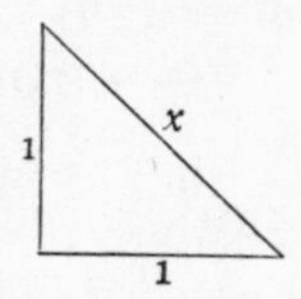

Then, $x^2 = 1^2+1^2 = 2$, or $x = \sqrt{2}$. The proof that this is irrational is then as above.

If we refer again to the correlation of numbers with points, we find that the above hypotenuse, since it is of definite length and can be carried off with a compass, may be transferred to this line of correlated points. Further, since x is not a rational number, then there appears to be a point which is correlated to $\sqrt{2}$, but is not correlated to any rational number. Since there are any number of irrationals, there must be any number of points correlated to these irrationals.[12] As a matter of fact, it can be demonstrated by means of the theory of aggregates (which we shall discuss later) that the rational numbers are the exceptional numbers, while the irrational numbers are met much more frequently on a line.

3. *Introduction of Irrational Numbers*: The introduction of irrational numbers into the number system is extremely difficult since, by their very nature, no finite set of rational or merely whole numbers can define them. Many ways have been offered of introducing these numbers, the best known being Dedekind's.[13] The discovery of the axioms for irrational numbers will not be considered. For such a set see Huntington's papers cited in the discussion of axiomatic definitions of numbers. We desire here rather to *develop* irrational numbers in terms of rational ones than to *postulate* their existence as independent entities. It is interesting to quote Russell here: "The method of 'postulating' what we want has many advantages; they are the same as the advantages of theft over honest toil."[14]

Two methods have been advanced for the introduction of irrational numbers: (1) Cantor's, (2) Dedekind's. These methods are the ones found in texts on mathematics, and are generally accepted as satisfactory.[15]

---

[12] Cf. Waismann, *op. cit.*, p. 8, and Hardy, *op. cit.*, pp. 4–5.

[13] Survey of these ways is in, *Encyclopédie des sciences mathématiques* Tome 1, Vol. 1, p. 133, Paris, 1904.

[14] Russell, *Introduction to Mathematical Philosophy*, p. 71. The reader should refer to Russell's discussion of irrational numbers, pp. 68 ff. Dedekind's definition of irrational numbers can be read in Dedekind's *Essays on Number*, Chicago-London, 1924, pp. 8–19. Also Hardy, *op. cit.*

[15] Cf. E. J. Townsend, *Functions of Real Variables*, N. Y., 1928, pp. 3–17. Cf. Russell's *Principles of Mathematics*, Chap. 34. Townsend shows these two to be equivalent, p. 12.

We turn to the introduction of irrational numbers.[16] A *bounded* sequence is one such that the absolute value of every term of the sequence is less than a given number. (By *absolute value* we mean the number irrespective of its sign.) By *sequence* we mean an ordered series. We shall denote the sequence $a_1 a_2 \ldots a_n \ldots$ by $\{a_n\}$.

Given a bounded sequence $\{a_n\}$ of rational numbers and a rational g, g is said to be the *limit* of $\{a_n\}$ if for any positive rational number k we can find a natural number h such that for every $i \geqq h$, $g-k<a<g+k$. We then say

$\{a_n\}$ *converges* to g. This criterion, however, assumes that we know g—the limit. It is possible, as is usually done, to give a criterion which can be derived from this condition and which enables us to determine whether or not a sequence is convergent even if we do not know its limit. This condition is: if any arbitrary $\epsilon > 0$ is given, there exists an N such that when $n > N$, $|a_n - a_{n+p}| < \epsilon$.

In general we shall in *monotone increasing* sequences take the smallest g which satisfies the inequality and in *monotone decreasing* sequences the largest g, as the limit.

A sequence is said to be monotone increasing (decreasing) if every term a_n is greater than (less than) its preceding term a_{n-1}.

If we realize that our definition of limits involves the existence of a *rational* number, then we see that sequences exist which do not have limits even though they may be bounded. Thus the sequence 1, 1.4, 1.41, 1.414, 1.4142,... does not exceed 1.5, yet we can find no *rational* number satisfying the conditions necessary for a limit. This sequence is actually the sequence of approximations obtained by extracting the square root of 2, and we have already shown that this root cannot be a rational number. At this point it is possible to do as Cantor did, postulate the existence of irrational numbers as limits of such sequence, or attempt to introduce them by other methods.

Dedekind was struck by the fact that every point on a line divides the points on the line into two classes of points. He then noted that any rational number a also divides the rational numbers into two classes such that all numbers in one class are less than a and all in the other are greater than a, while a can be conceived as being either the last element in one or the first term in the other class. The rational number a might be viewed as being defined by this *cut* or separation of the number system.

It is possible to postulate a limit to the sequence 1., 1.4, 1.41, etc., to show this to be consistent with the rest of the number system, and to define the operations with these numbers. It is difficult to see why this method is in itself to be discarded. This would really mean that

---

[16] In what follows we are indebted to Kaufmann, *op. cit.*

every bounded sequence has a limit and where the limit is not a rational number we have a new type of number, the irrational.

But to return to Dedekind's cut. We define a *cut* as follows:

If the set of rational numbers M is divided into two sets, M_1 and M_2 such that (1) every number M belongs to one and only one subset M_1 or M_2; (2) neither M_1 nor M_2 is an empty set; (3) every number in M_1 is less than every number in M_2; then this division is said to be a *cut* in M. We denote the cut by $(M_1/M_2.)$

Now four cases are possible if we cut the rational number series.

1. M_1 has a last (greatest) and M_2 a first (least) number.

2. M_1 has a greatest but M_2 has no least number.

3. M_1 has no greatest but M_2 has a least.

4. M_1 has no greatest and M_2 has no least.

If the cut is such that Case 1 occurs we call it a *bound* (*Sprung*). *Cuts that are bounds define whole numbers.*

If the cut is such that Case 4 occurs, we call it a *gap* (*Lücke*).

Cases 2 and 3 are called *continuous cuts.* When Cases 2 and 3 occur we have rational numbers. For instance, let us take 1/2. In M_1 put all numbers less than 1/2, in M_2 all numbers greater than 1/2.

1/2 can be put either in M_1 as its last term (in which event we have Case 2), or in M_2 as its first term (in which event we have Case 3).

Now there are sequences (as, e.g., that for $\sqrt{2}$) which no rational numbers generate. We are therefore led to extend the number system to include these *irrational* numbers. The irrational number can therefore be defined as a cut in the rational numbers and therefore in terms of the natural number system—Case 4.

It is possible to look at the irrational number, in a slightly different fashion, as the left hand *class* of the cut. In this way we need not consider the limits of either the lower or upper class but treat the class itself.[17]

The series of rational and irrational numbers is called the series of *real* numbers.

(It can be demonstrated that every segment of the sequence of real numbers defines another real number.)[18]

This change in method of defining number, i.e., in terms of a sequence of numbers, is a fundamental change of point of view. The method will enable us to define another number of a different type—the *transfinite* number.

---

[17] This is Russell's method. Cf. *Introduction to Mathematical Philosophy*, pp. 21–23, and *Principles of Mathematics*, Chaps. 33, 34, but Cf. Kaufmann, *op. cit.*, p. 127. p. 127.

[18] Cf. Townsend, *op. cit.*, pp. 15–16.

(Kaufmann insists that irrational numbers as roots of an algebraic equation are merely symbols which denote that we have a law of construction for any rational approximation to the roots of an equation. He feels that in this way the necessity of an infinite set of rational numbers in the definition of irrationals is avoided.[19] It is, however, difficult to understand what an approximation is which has nothing to approximate. That to which these are approximations are the irrational numbers and hence Kaufmann falls back to Dedekind's definition.)

Our discussion so far has been an attempt to construct the various types of numbers from the natural (i.e., the positive whole) number series. Up to this point all the numbers we found were also seen to be the roots of algebraic equations. This is not true of all irrational numbers. $\sqrt{2}$, $\sqrt{3}, \ldots$ are roots of such equations, e.g., $x^2-2 = 0$. But the ratio of the diameter of a circle to the circumference (i.e., π) and the base of natural logarithms (i.e., e) can be demonstrated not to be roots of algebraic equations. An algebraic equation is an equation of the form $a_0x^n+a_1x^{n-1}+a_2x^{n-2}+\ldots+a_{n-1}x+a_n = 0$, where all a_i are rational numbers. Numbers which are roots of algebraic equations are called algebraic numbers, all others are called *transcendental*. This division of the number systems is useful for certain purposes but is not of importance here. It is possible to show that there exists an infinite number of transcendental numbers. This we shall do later in connection with our discussion of the theory of aggregates. All transcendental numbers are irrational and can therefore be defined either as a cut or as the lower class of a cut. The series which define e and π can be found in any elementary calculus text.[20]

Irrational numbers are not definable by means of rational numbers but only by *classes* of rational numbers.[21] In other words, an irrational number is a class composed of classes of classes similar to each other (which can, that is, be put into one-to-one correspondence). If we consider a class whose elements are individuals to be of Type 1, and a class whose elements are classes of Type 1 to be of Type 2, and a class whose elements are classes of Types 2 to be of Type 3, then irrational numbers are defined in terms of classes of Type 3 while rational numbers are defined in terms of classes of Type 2. If classes of Type 3 are not reducible to classes of Type 2, then irrational numbers are not reducible to rational numbers.

It would appear that if we defined a class of classes of Type 3, we ought to get a new type of number. But the theorem that was referred to as proven in Townsend, that any class of classes of Type 3 con-

[19] Kaufmann, *op. cit.*, pp. 128–29.

[20] F. Klein, *Elementary Mathematics From an Advanced Standpoint*, pp. 237–50, for the proof of the transcendental character of e and π.

[21] Cf. Waismann, *op. cit.*, pp. 167–68.

verges to a real number, seems to indicate that this would result in a class of Type 2 or Type 3.[22] This really means that we have reached a kind of barrier to number-constructions in terms of classes of numbers.

COMPLEX NUMBERS

The irrational number results from an equation but is definable as a class of rational numbers. It is recalled also that fractions can be defined as classes of two numbers, i.e., as couples whose operations are definable. The next type of number will also be definable in terms of couples whose operations are defined differently from those of the fractions. These numbers cannot be defined in terms of sequences of rational numbers and are therefore not real numbers. Unfortunately the word "imaginary" was given to these new non-real numbers, with the result that a great many pseudo-problems were bitterly disputed.[23] The discovery of a geometric interpretation for these showed them not to be completely imaginary.[24]

1. *Algebraic Discovery*:[25] Consider the equation:

$$x^2+1 = 0. \text{ Then } x^2 = -1 \text{ and } x = \sqrt{-1}.$$

A negative number is never the square of any real number. It appears then that either we have to limit the operation of extracting roots, or say that such equations are inconsistent, or try to set up a new type of number. The desire to treat $\sqrt{-1}$ (denoted by i) as a new type of number is increased when we note that by its use we can relate functions of angles and the transcendental number e. Euler's equation, $e^{ix} = \cos x + i\sin x$, does precisely that.

Consider further the equation $x^2+2x+2 = 0$; we know that the general solution of $ax^2+bx+c = 0$ is given by $x = \dfrac{-b\pm\sqrt{b^2-4ac}}{2a}$

Substituting, we get

$$x = \frac{-2\pm\sqrt{4-8}}{2} = \frac{-2\pm\sqrt{-8}}{2} = \frac{-2\pm2\sqrt{-1}}{2} = \frac{-1\pm\sqrt{-1}}{} .$$

[22] The theory back of the theory of types was developed by Russell to avoid certain paradoxes which we shall discuss later. He also introduced the *axiom of reducibility*, which says that any statement in terms of a given type can be replaced by an equivalent statement in terms of a lower type. The difficulties in these ideas and the attempts to rid the theory of logic of a theory of types is beyond our scope here. Cf. Black, *Nature of Mathematics*, pp. 101–18.

[23] Such phrases as made by Leibniz in 1702 (quoted by F. Klein, *Elementary Mathematics From a Higher Standpoint*, trans. p. 56) that "imaginary numbers are a fine and wonderful refuge of the divine spirit, almost an amphibian between being and non-being" were the result.

[24] For the development during the nineteenth century, *see* Stammler, *op. cit.*, pp. 32 ff.

[25] Cf. the discussion in Hölder, *Mathematische Methode*, pp. 199–204, and Shaw, *Philosophy of Mathematics*, Chap. 7.

This is known as a *complex* number. In general any number of the form $a+bi$, where a and b are real, is a complex number. ($\sqrt{-2}$ is of the form, $0+\sqrt{-2}$. A real number can be considered as $a+0i$.)[26]

2. *Geometric Discovery*: The complex number was not discovered geometrically. Rather it took some years to discover a geometric interpretation for these complex numbers.

3. *Arithmetic Constructions*: Complex numbers may be regarded as couples of real numbers in the same way that fractions were. The rules of operation are of course different from those for fractions.

$(a,b) = (c,d)$ if $a = b$ and $c = d$; e.g., $3+5i = 3+5i$ if $3 = 3$ and $5 = 5$
$(a,b) \pm (c,d) = (a \pm c, b \pm d)$; e.g., $(3+5i)+(6+7i) = (3+6)+(5+7)i$
$(a,b) . (c,d) = (ac-bd, ad+bc)$.

We shall not discuss the higher forms of complex numbers, except to point out that there are such.

TRANSFINITE NUMBERS

The discovery of new types of numbers has taken place as a result of the necessity for making solutions of algebraic equations possible. This has led us a great way. But we seem to have come to the end of that path. Are there other paths possible? If so, can we find any other type of number?[27]

Let us recall again that the natural number series was defined in terms of classes of type n (n finite). It was possible to exhaust all the elements satisfying $f(x)$. In other words, $(x)f(x)$ defined a set having a last term. Consider now a class of objects which is non-finite in the sense that it has no last term, e.g., the class of finite integers 0, 1, 2, . . ., or the class of points on a line one inch long but not including the last point of the segment. How many objects do these contain? We must try to free our mind here of deep-rooted habits of thought. "How many" does not mean a finite number; obviously no finite n tells how many objects these classes have. Recall that for Russell a number is a class of classes similar to each other; i.e., a class of classes which are (1,1) correlatable to each other. This means that if we have classes of what is said to be a non-finite number of elements which can be correlated (1,1) to each other, then the class of these forms a new number

[26] Cf. Russell, *Introd.*, pp. 75–76. Hardy, *op. cit.*, pp. 80–81, introduces complex numbers as symbols for displacement in a plane.

[27] As a matter of fact, the work of Gödel, which we shall discuss later, seems to indicate that the process of constructing new real numbers is unending. He seems to demonstrate that it is possible to construct real numbers within any arithmetic which are not definable within that arithmetic, i.e., it is always necessary to construct a wider arithmetic. This implies a relation between definability and constructability which may be of importance in discussion of the Hilbert-Brouwer controversy but is beyond our scope here. Cf. M. Black, *op. cit.*, pp. 167–68. Cf. R. Carnap, *Logical Syntax of Language*, N. Y., 1937, p. 220.

called the *transfinite* number. There are certain apparently rather paradoxical properties of these numbers, but they appear to be paradoxical only because we continue to think of them as though they were finite numbers.

Another approach to these transfinite numbers is in the process of thought used to define the irrational numbers. Recall that the number was defined there as a class of rational numbers. Consider the rational number series 1, 2, 3, 4,.... This is a well defined series. As such it can define a number which is not one of its elements. This number which the series of whole numbers defines is called a transfinite number.

There seem to be different transfinite numbers and rules of operation, for them can be uniquely defined.[28]

The theory of these transfinite numbers will be developed in the next chapter on the theory of aggregates (classes). The importance of the theory of aggregates cannot be underestimated, since we can see now how the entire structure of mathematics seems to be based upon them. As a matter of fact, the theory of real variables is now built upon aggregates of points.[29]

With the introduction of transfinite numbers, many new problems arise. Are there really infinite classes? If not, then there exist only finite classes and therefore only finite numbers. Classes such as the number series would then be called *potentially* infinite, by which is meant that it is always possible to add an element after the last one we have reached. But if that is so, since such a class would be different from one in which that were not possible, it ought to have correlated with it classes of similar kind and therefore gives rise to a new number. In such a case whether we have an actual or only a potential infinite would really be irrelevant to mathematics.

Are there kinds of transfinite number? Can we give a precise definition to *infinite class*?

CONCLUDING REMARKS TO EXTENSION OF NUMBER SYSTEM[30]

Our entire discussion has been in terms of cardinal numbers; numbers obtained by correlating the elements of classes. Nothing has been said about the *order* which exists in the various series of numbers. So far as the finite numbers are concerned, the number which expresses the order of the elements is the same as that of the cardinal number. In

---

[28] The foundations of the theory of transfinite numbers were laid by G. Cantor, a German mathematician, in 1895. This paper has been translated by Jourdain and published in English. Cantor, *Transfinite Numbers*, Chicago, and London, 1915. It has an excellent introduction by the translator.

[29] Cf. Townsend, *op. cit.*, Chap. 2.

[30] The student should consult Waismann, Chaps. 2 and 16, and also Shaw, *op. cit.*, Chap. 2.

transfinite numbers, as we shall see, ordinal numbers express the *type* of order and are not the same as cardinal numbers. We shall conclude this chapter with three sets of postulates describing the type of order of (1) the natural number series; (2) the series of rational numbers; (3) the series of real numbers. These are taken from Huntington, *The Continuum*, Chapters III, IV, and V. (Huntington seems to treat the numbers as mere tags to the elements of the series.)[31]

All types of order have the following postulates in common:

Given a class K and a relation $<$,—i.e., the system $(K, <)$.

Postulate 0. K is not empty nor a class containing only one element.

1) If a and b are distinct elements of K, then either $a < b$ or $b < a$.

2) If $a < b$, then a and b are distinct.

3) If $a < b$ and $b < c$ then $a < c$.

Discrete series have in addition:

(N 1) (Dedekind's Postulate) If K_1 and K_2 are any two non-empty parts of K, such that every element of K belongs either to K_1 or K_2 and every element of K_1 precedes every element of K_2, then there is at least one element x in K such that

(1) any element that precedes x is in K_1, and

(2) any element that follows x is in K_2.

(N 2) Every element of K except the last has an immediate successor.

(N 3) Every element of K except the first has an immediate predecessor.

The natural numbers form a *discrete* series.

Dense series have postulates 0–3 and in addition:

(H 1) If a and b are elements of K and $a < b$, then there is at least one element x in K such that $a < x$ and $x < b$.

(H 2) The class K can be put into one-one correlation with a discrete series having no last element (i.e., K is *countable*). Such a series is 1, 2, 3, 4,. . . .

(This process of correlation with such a discrete series is really the foundation of counting.)

The rational numbers form a *dense* series.

Continuous series have postulates 0–3 and in addition N-1, of the discrete series, H-1 of the dense series, and for *linear continuous* series also:

C 3) The class K contains a countable subclass R in such a way that between any two elements of K there is an element of R.

The real numbers form a continuous series. The essence of the problem of continuity lies in the type of order of continuous series.

---

[31] E. V. Huntington, *The Continuum*, p. 25. Cf. for a discussion of order, Russell, *Principles of Mathematics*, pp. 199–227.

APPENDIX

That the extension of the number system as we have outlined it is not completely satisfactory is almost self-evident. At each stage we are compelled to introduce what amounted to new theories. Instead of a succession of either generalizations or specializations we described a succession of *theories*. In other words, *we do not have a real deductive theory of numbers*. The inadequacy of the extension is felt at such places as complex numbers and irrationals when they are treated as couples. This idea is emphasized by Padoa in an article read to the International Congress of Scientific Philosophy at Paris in 1935, and published in the *Actes du Congrès International de Philosophie Scientifique*, VII, Logique, Paris, 1936, p. 53, under the title "Les extentions successives de l'ensemble des nombres au point de vue déductif." Despite some weaknesses at particular points of Padoa's argument, his position is correct in principle.

Padoa points out that in order to construct *deductively* an arithmetic theory, it is necessary to pre-establish the widest types of numbers that one intends to use and to introduce this as a primitive idea (p. 56). Padoa claims to have done this, i.e., have constructed a deductively unified arithmetic containing all numbers but the transfinite ones (whose existence he denies). Unfortunately, this is not available to the author and therefore can only be mentioned here.

CHAPTER SEVEN

A. THEORY OF AGGREGATES AND TRANSFINITE NUMBERS[1]

It is clear now how important the notion of a class (aggregate) is. All types of number depend fundamentally upon this concept. The possibility of transfinite numbers also depends on this concept. It was in connection with the discussion of transfinite numbers, as a matter of fact, that Cantor laid the foundations for a theory of aggregates. Not that the ideas were not already present. Bolzano had done some work in connection with the concept of infinite aggregates. Actually in his *Paradoxien des Unendlichen* (1851) he gave what turned out be be the definition of an infinite set. (Unfortunately no translation in English of this important work exists.) Also, of course, the development of the logic of classes in symbolic logic by Frege, Boole, etc., went hand in hand with the development of Mengenlehre.[2] Involved then in the theory of aggregates is the entire structure of mathematics and the approach to a discussion of the foundations. This theory may be termed psychological also, since it involves (1) ability to group objects, (2) ability to abstract from the nature of objects.

1. THEORY OF AGGREGATES

An aggregate is defined as a unification of distinct well-defined objects of intuition or thought into a whole. These objects are called the *elements* of the aggregate. An aggregate, says Hausdorff, is a plurality conceived as unity. Thus any objects (e.g., pear, pencil, paper, picture) when viewed as forming a unit (i.e., the collection of these objects) form an aggregate. Examples of aggregates are (1) the aggregate of algebraic numbers, (2) the aggregate of points on a line. If we consider the elements $a,b,c\ldots$, then the aggregate of these will be de-

[1] A popular discussion of the elements of the theory of aggregates is to be found in T. Dantzig, *Number, the Language of Science*, Chap. 11. The best elementary introduction is A. Fraenkel, *Enleitung in die Mengenlehre*, 3rd ed., 1928, Berlin, which includes a most comprehensive bibliography on philosophy of mathematics.

[2] The distinction between classes and aggregates is beyond our scope. But cf. Russell, *Principles*, Chap. 6 and Chap. 7. The following outline of the theory will follow Fraenkel closely. An almost too simple exposition of much of what follows is in Tarski, *Einführung in die Mathematische Denken*, Vienna, 1937. This is now available in English as *Introduction to Logic*, New York, 1941.

noted by $M = \{a,b,c \ldots\}$. Thus the aggregate of rational numbers is $\{1,2,3 \ldots\}$. An aggregate is more than its elements and must never be viewed as being composed of its elements as a body is of its parts. An aggregate is an *abstraction*. From this point of view we must distinguish, as Peano did, the entity "a" from the aggregate composed of one element a, i.e., $\{a\}$. Thus *Socrates* is not the same as the class whose only element is Socrates.

By the term *distinct* we mean that for any two objects in question we shall know whether they are to be considered the same or different. By *well-defined* we mean that we can determine whether or not the term belongs to the aggregate in question.

Two aggregates are said to be equal ($M = N$) if they contain the same elements.

Definition 1: Two aggregates M and N are said to be equivalent if the elements of M can be bi-univocally correlated with the elements of N. A bi-univocal correlation is a correlation such that to every element of M there corresponds one and only one element of N and to every element of N there corresponds one and only one element of M.

Any correlation can be set up by means of a function. In fact every function is a correlation between two aggregates. $y = f(x)$ denotes that y is correlated with x by means of f. Any bi-univocal correlation ϕ between two aggregates will be called an *image*. It therefore follows that two aggregates are equivalent if there exists an image between them. Since every aggregate can be imaged to itself, every aggregate is equivalent to itself. We shall use the symbol $\sim$ to stand for equivalence.

Theorems (to be proved by reader):

1. The relation of equivalence is symmetric, i.e., if $M \sim N$, $N \sim M$.

2. The relation of equivalence is transitive, i.e., if $M \sim N$ and $N \sim R$, then $M \sim R$.

Definition 2: An aggregate M' is said to be a *subaggregate* of M if every element of M' is also an element of M. If every element of M' is an element of M but not conversely, then M' is said to be a *proper* subaggregate of M.

We shall consider the null aggregate, i.e., the aggregate which contains no elements and is therefore not properly speaking an aggregate, to be an aggregate and also to be a subaggregate of every aggregate. This is needed for simplification of various theorems. It is denoted by 0—to be distinguished from $\{0\}$, and $\{\{0\}\}$, etc.

The introduction of the null aggregate as a subaggregate of every aggregate enables us to set up the theorem that if n is the number of elements in the aggregate (n finite) then the number of subaggregates will be 2^n, e.g., $\{1,2,3,4\}$ has $\{0\}$, $\{1\}$, $\{2\}$, $\{3\}$, $\{4\}$, $\{1,2\}$, $\{1,3\}$, $\{1,4\}$, $\{2,3\}$, $\{2,4\}$, $\{3,4\}$, $\{1,2,3\}$, $\{1,2,4\}$, $\{1,3,4\}$, $\{2,3,4\}$, $\{1,2,3,4\}$; i.e., $2^4 = 16$.

Definition 3: If M' is a proper subaggregate of M and $M' \sim M$, then M is said to be a *transfinite aggregate*.

Bolzano had called this a *paradox of the infinite*. It is paradoxical only because of our ordinary ways of thinking. In a certain sense this means that the whole is not "greater" than its part so far as transfinite aggregates are concerned. It does not mean that the whole is never greater than its part. When we count objects we really set up a correlation between the objects and the natural number series, e.g.:

Apple	Apple	Apple
↕	↕	↕
1	2	3

The last number used tells us "how many" there are. If we compare two groups we do not need the numbers. Suppose we have an aggregate of apples and one of pears. Then, if every time we throw out an apple we also throw out a pear and when we have no more apples we have no more pears, then we say we have as many apples as pears. (This also indicates how number is a property of aggregates.)

Consider the natural number series:

$$N = 1, 2, 3, 4, 5, 6, 7, 8, \ldots$$

and the series of even numbers:

$$N' = 2, 4, 6, 8, 10, 12, 14, 16, \ldots.$$

The image is $n = 2n_1$ (and conversely $n_1 = \frac{1}{2}n$). Therefore, $N \sim N'$ and N' is a proper subaggregate of N. Therefore, N is transfinite (being imaged on a proper subaggregate of itself).

Note that this definition of *transfinite* is a positive one, and finite aggregate is defined negatively—as one which has no proper part equivalent to itself.

Definition 4: If M can be imaged on the natural number series, then M is said to be *countable*.

This means that we can order the elements of the aggregate. A countable aggregate is always infinite although in ordinary usage we speak of finite aggregates as countable and distinguish between countable finite and countable infinite aggregates.

(Is every infinite subaggregate of the aggregate of natural numbers countable? Can you prove that every infinite subaggregate of a countable aggregate is countable? Can you count the aggregate of positive and negative whole numbers?)

As illustrations of countable aggregate we shall consider the aggregates of rational numbers and of algebraic numbers.

We'll consider the aggregate of rational numbers, i.e., of numbers that can be written in the form m/n. Consider first the case where m and n are positive. Let $S = m+n$. If we have S given, then obviously other sets of numbers exist which add to S, e.g., if we have 2/3, then $S = 2+3 = 5$; if $S = 5$ is given, then S may be $4 + 1$, or $3 + 2$, or $2 + 3$, or $1 + 4$. (All whole numbers will be of the form $m/1$ and $0 = 0/1$.) We shall consider for any given S all those whole numbers which add to S, but have no common factors. For a given S we shall arrange the possible fractions as follows: (1) Put first that fraction having the greatest numerator and smallest denominator, then arrange the fractions in order of decreasing numerator and increasing denominator. E.g., if $S = 7$, then we have $6 + 1$, $5 + 2$, $4 + 3$, $3 + 4$, $2 + 5$, $1 + 6$, or in order 6/1, 5/2, 4/3, 3/4, 2/5, 1/6. Since S is finite, there will be only a finite number of pairs adding to S. We now are ready to order the rational numbers and thereby count them. First we put $0 = 0/1$, for $S = 0+1 = 1$. Then we put all fractions for which $S = 2$, arranged as described above, i.e., 1/1. Then those for which $S = 3,4 \ldots$ We insert all negative numbers as follows: after each rational number m/n, put $-m/n$. We thereby have the sequence of rational numbers beginning as follows:

$$0/1,\ 1/1,\ -1/1,\ 2/1,\ -2/1,\ 1/2,\ -1/2,\ 3/1,\ -3/1,\ 1/3,\ -1/3,\ 4/1,\ -4/1,\ 3/2,\ -3/2,\ 2/3,\ -2/3 \ldots$$

The position of *any* rational number can then be found and the series is thereby counted as follows:

If p/q is a positive number, then $t = p+q$, which is finite and has a finite number of rational numbers such that $m+n = s < t$. Therefore, after a finite number of steps we can count up to the given number p/q. If p/q is negative we find it one place beyond its positive position. This procedure enables us to image the rational numbers on the whole numbers. (The rational numbers can be imaged to the rational points on a line and thereby show the equivalence of the aggregate of rational numbers with the aggregate of rational points.[3])

[3] Cf. E. W. Hobson, *Theory of functions of a real variable*, Vol. 1, pp. 80–81.

Consider the real algebraic numbers.[4] These were defined as the real roots of equations of the form

$$a_n x^n + a_{n-1} x^{n-1} + \ldots + a_2 x^2 + a_1 x + a_0 = 0$$

where a_i are all whole numbers and $a_n \neq 0$. Every rational number is of course an algebraic number (as a solution of $qx - p = 0$). $|a_i|$ = *absolute* value of a_i.

Now let $h = (n-1) + |a_n| + |a_{n-1}| + \ldots + |a_2| + |a_1| + |a_0|$, i.e., $h = (n-1) + \sum_{i=1}^{n} |a_i|$. Call h the *rank* of the equation. Then clearly *every* equation has a rank. But for any given h there are only a finite number of equations of rank h possible. Now we can order all algebraic equations as follows: first come all equations of rank 1, then those of rank 2, then those of rank 3, etc. As for those equations of equal rank, any order may be assigned them; e.g., an order according to degree. We can thus determine the correlated whole number for *any* equation. Any equation has, at most, n distinct roots. Therefore each equation has a finite number of roots. We can, therefore, order the roots of an equation, say, according to magnitude. We lay down the condition that a root of an equation is taken only the *first* time it occurs. If it occurs again we neglect it. The usual order of numbers is obviously different—but we do have an order here. Given an algebraic number, then we can determine the *rank* of its possible equation and determine where in the order of equations this occurs. We can then set up the sequence of algebraic numbers that precede this—and since there are only a finite number, count this array.

(The reader should note carefully how handling the infinite is reduced to a process of handling *any* entity of the given type. The infinite seems to be based on the idea of the presence of a next after any one. The distinction between an actual infinite and a potential one is based on this idea.)

It is possible to show the existence of an infinite number of the non-algebraic (=transcendental) numbers—which we mentioned before. We first recall that every real number can be expressed as an infinite decimal fraction (in the case of whole numbers we can add an infinite number of 0's after the decimal point. Obviously a similar device is in order for rational numbers whose denominators are positive powers of 10.)[5] Consider the real algebraic numbers. These can

---

[4] If we allow infinite recurrences of the same number (e.g., $1 = 2/2 = 3/3 =$) then the positive rational numbers positions can be computed: if p/q then its position is $1/2\ (p+q-1)(p+q-2)+q$. Cf. Hardy, *Course of Pure Mathematics*, p. 1.

[5] Cf. Hardy, *op. cit.*, pp. 149–51.

be ordered as described above. We write down after each number its corresponding decimal fraction as follows:

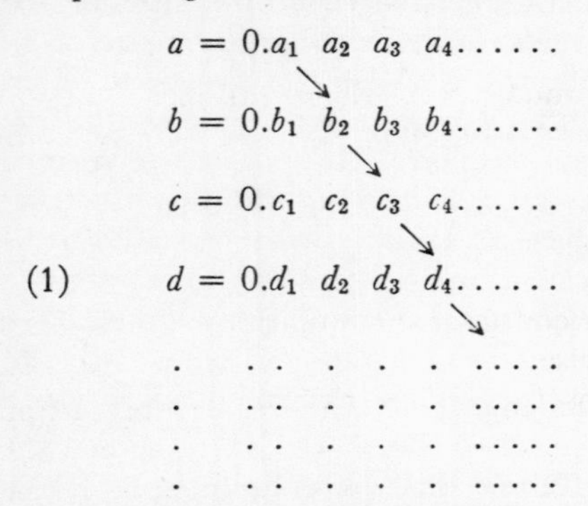

Let (1) represent the counting of real algebraic numbers, where $a_i, b_i, c_i, \ldots$ represent digits 0 to 9.

We are to construct a number not in this series and, therefore, not an algebraic number. Consider the first number a. If a_1 is 0, we shall write down 1, if a_1 is any other number, we write down 0. Consider b—if b_2 is 0 we write down 1, otherwise 0; consider c—if c_3 is 0 we write down 1, otherwise 0, etc. In general consider the nth number. If its n'th decimal place is 0, then in the number we are constructing we write in the nth place 1, otherwise 0. We then have a number $0.\delta_1\delta_2\delta_3\delta_4\ldots$ which we have constructed on this principle and which differs from the nth real algebraic number at least in the nth place. Since the array constructed had every real algebraic number, this number is not in the array and is therefore not algebraic. Obviously this procedure can be extended so as to allow the construction of any number of transcendental numbers.

A great deal of controversy centers around the use of such transfinite methods of procedure. We know two transcendental numbers even if we cannot use this method—known as the *diagonal method*—to construct them, i.e., e and π. But we have not been able to determine other transcendentals by this method. The diagonal process allows the theoretical demonstration of the possibility of such numbers. Here the whole question of existence in mathematics arises. Have we really demonstrated the *existence* of these numbers or must we construct them first? In what sense have we constructed π and not constructed the number $0.\delta_1\delta_2\delta_3\ldots$? The above proof is based on the *actual* counting of the algebraic numbers, which is itself based on a *law* of counting them. We seem to be involved in giving a law in general terms for constructing numbers and this law itself is based on a law whose nature is like its own.

We have shown that the aggregate of algebraic numbers is countable. Is the aggregate of all real numbers also countable? Obviously if all infinite aggregates are equivalent we have very little clue for the construction of a theory of transfinite numbers. The aggregate of all real numbers, for very obvious reasons, is called the *continuum*. That the continuum is not countable can be demonstrated by using a *reductio ad absurdum* proof and the diagonal method. The point is to assume the real numbers to be countable, and count them, and then construct a number not in the counting. Since all numbers were assumed to be in the counting, the construction of a number not in the counting is obviously absurd. Therefore, the continuum cannot be counted. We leave the actual proof as an exercise.

So much for these illustrations (which are extremely important). We return to our development.

Two aggregates which are equivalent obviously have a certain property in common. This property common to all equivalent aggregates is called the *potency* of the aggregate. In the case of finite aggregates, it is clearly the cardinal number. (Cf. the development of Russell, Frege, *et al.*) The fact that the continuum is non-countable and therefore not equivalent to the aggregate of (say) algebraic numbers indicates the possibility of different potencies. "The cardinal number of an aggregate M can be viewed as the totality of those properties of M which M has in common with every aggregate equivalent to it but are not present in any non-equivalent aggregate."[6] (Since a totality of properties defines a class, this definition is not essentially different from Russell's.) We can, therefore, say that if two infinite aggregates possess the same potency they are equivalent, and that if the potencies of two aggregates are not equal, the aggregates are not equivalent.

The next step is clearly to introduce a set of symbols for these new cardinal numbers. Cantor used the Hebrew alphabet with subscripts: $\aleph_0, \aleph_1, \aleph_2, \ldots$ (read "aleph-null," "aleph-sub-one," etc.). We shall, however, temporarily use small Latin letters. We shall let a be the potency of every countable aggregate and c the potency of the aggregate of real numbers. (The aggregate of all possible functions $f(x)$ of x has a cardinal number not equal either to a or c. Cf. Proof in Fraenkel, pp. 61–63.)

We shall now introduce the fundamental rules of operation for transfinite cardinal numbers.

Definition 5: If the aggregate M is equivalent to a subaggregate of N but N is not equivalent to a subaggregate of M, then the cardinal number of M is less than the cardinal of N. (If m is cardinal of M and n of N, then we write $m < n$.)

[6] Fraenkel, *op. cit.*, p. 57. *See also* Cantor, *Transfinite Numbers* (trans. by Jourdain), Chicago, 1915, pp. 86 ff. Jourdain translates "Mächtigkeit" as power; we use potency. Cf. also Hobson, *op. cit.*, Chap. 4.

This definition refers the relation of magnitude among transfinite cardinals back to the concept of equivalence and subaggregate. On the basis of this definition, we can show that if $m < n$, n cannot be less than m; if $m < n$ and $n < p$ then $m < p$, etc. The reader should try to prove that (1) $a \leq c \leq f, f$ being the cardinal of the aggregate of functions. (2) a is smallest transfinite cardinal.

Definition 6: By the *sum* or *sum aggregate* of two aggregates M and N we mean the aggregate S containing all elements which occur either in M or in N (i.e., at least in one of the two aggregates). We write $S = M+N$ or $S=\Sigma(M,N)$. (If an element occurs in both M and N, it is only taken once in S.)

(Definition 6 can be extended to include any number of aggregates: If $M = \{M_1M_2M_3\ldots\}$,

then $S = \Sigma\{M_1M_2M_3\ldots\} = M_1+M_2+M_3+\ldots$).

Definition 7: By the sum of the two given cardinal numbers of M and N we mean the cardinal number of $M+N$, i.e., if m is cardinal of M, n that of N, and s that of $M+N$, then $m+n = s$. E.g., if $M = \{1,2,3\}$, with cardinal number 3; and $N = \{4,5\}$, with cardinal number 2; then $3+2$ is the cardinal of $M+N = \{1,2,3, 4,5\}$ which is 5.

It is now possible to prove that $a+b = b+a$ and that $(a+b)+c = a+(b+c)$.

(The reader should try to show that, if a is the cardinal number of the class of natural numbers and n the cardinal number of any finite class of natural numbers, then $n+a = a$.)

Definition 8: If an element m of the aggregate M be associated with an element n of N so as to form the couple (m,n), then the aggregate of all possible elements that can be found in this way is called the *product* of M and N.[7]

We write $P = M \cdot N$. E.g., if $M = \{1,2,3\}$ and $N = \{4,5\}$, then $M \cdot N = \{(1,4), (1,5), (2,4), (2,5), (3,4), (3,5)\}$.

Definition 9: The product of the given cardinals of M and N is the cardinal of the product of M and N; i.e., $m \cdot n = p$ where p is cardinal of $P = M \cdot N$.

We shall not carry this development further. The interested reader can begin with the chapter in Hobson already referred to and also Cantor's paper translated by Jourdain.

So far we have been dealing with cardinal numbers and we now briefly consider ordinals. As we remarked above, the ordinal differs from the cardinal so far as transfinite numbers are concerned.

If now we consider two equivalent aggregates, paying no attention to the *kind* of elements they contain, then we note they may differ in what we call the arrangement or ordering of their elements. If we consider, e.g., M as the aggregate of natural numbers and N as the aggregate of fractions in order of magnitude, then we can rearrange N

[7] Hobson, Vol. 1, *op. cit.* pp. 192–93.

and show $M \sim N$, but in this order M has a first element and N does not. We can of course rearrange a given aggregate so that although it contains the same elements it contains them in a different order, e.g.,

$\{\ldots -4, -3, -2, -1, 0, 1, 2, 3, 4, \ldots\}$ and

$\{0, 1, -1, 2, -2, 3, -3 \ldots\}$.

Definition 10: An aggregate M is said to be *simply ordered* if there exists a law which determines for any pair $\{a,b\}$ of elements of M that $a < b$ or $b < a$. ($<$ = precedes.) This law must be such that (1) $a < a$ is never possible (i.e., it is non-reflexive); (2) if $a < b$ then $b < a$ is impossible (i.e., asymmetric); (3) if $a < b$ and $b < c$ then $a < c$ (i.e., transitive).

There are often a number of different ways of ordering a given aggregate; they correspond to different choices among possible definitions of "precedes."

Definition 11: Two ordered aggregates M and N are said to be *similar* ($M \simeq N$) if we can correlate the elements of M to those of N in such fashion that (1) *every* element of M is *bi-univocally* correlated with *every* element of N, and (2) if m is correlated to n and m^1 to n^1 and if in $M, m < m^1$, then $n < n^1$ in N— and conversely.

It follows of course that if $M \simeq N$, then $M \sim N$, but not necessarily the converse. Also $M \simeq M$; and if $M \simeq N$, then $N \simeq M$; and if $M \simeq N$ and $N \simeq P$, then $M \simeq P$.

If we consider aggregates similar to each other, then that property which they have in common which makes them similar will be called the *order type* of the aggregates. If therefore we say that two aggregates have the same order type, we mean that they are similar. It will also mean that they have the same cardinal number. From this point of view the concept of *similarity* is wider than that of *equivalence.*

The order type of countable infinite aggregates is denoted by ω. The order type of the inverse of the countable aggregate (i.e., the aggregate $\{\ldots 4, 3, 2, 1\}$ which has no first but a last) is denoted by $^*\omega$. Not all order types are ordinal numbers, even though we can add and multiply them.[8] The addition of order types can be defined in terms of the aggregates whose order type they are. If μ is the order type of M and ν of N, then, if we form $S = M+N$— where in S the order of elements in M and N are kept and every element of M precedes every element of N, then the order type of S is said to be the order type of $M+N$, i.e., $\sigma = \mu+\nu$. (Note that $\mu+\nu \neq \nu+\mu$, e.g., $\{1, 2, 3, \ldots\}+\{-1, -2, -3\} = \{1, 2, 3, \ldots, -1, -2, -3\}$ while $\{-1, -2, -3\}+\{1, 2, 3, \ldots\} = \{-1, -2, -3, 1, 2, 3, \ldots\}$ the first sum has a last term the second does not.)

Prove $\omega+n \neq \omega$ but $n+\omega = \omega$; ω is the order type of countable aggregates. Show $\omega+1 \neq \omega+2 \neq \omega+3 \neq \ldots \neq \omega+n$.

[8] Since ordinal numbers are order types of well-ordered aggregates and there are aggregates which are not well-ordered, there are order types which are not ordinal numbers. There cannot be ordinal numbers which are not order types.

Show that $^*\omega+n = {}^*\omega$ but $n+{}^*\omega \not\simeq {}^*\omega$. Show that $\omega+{}^*\omega \not\simeq \omega$ and $\not\simeq {}^*\omega$.

Multiplication of order types can also be defined as follows:

If μ is the order type of M and v that of N, then $\mu . v$ is the order type of $M \cdot N$. (Again $\mu . v \not\simeq v . \mu$.) As in the case of addition, the multiplication of order types is referred back to the multiplication of aggregates. In order to multiply the order types of two aggregates, we multiply the aggregates. The order type of the product of the two aggregates is, by definition, the product of the order types of the given aggregates.

Definition 12: An ordered aggregate M is said to be *well ordered* if every subaggregate of M different from the null-aggregate has a *first* element.[9]

Every *counted* aggregate is well ordered, but not every *countable* aggregate.

We shall call the order types of these well ordered aggregates *ordinal numbers.* The term is restricted to these order types because of their greater likeness to numbers as we know them.

Definition 13: A subaggregate A of the well ordered aggregate M is called an *initial part* of M if, when any arbitrary element a belongs to A, every element in M preceding a belongs to A. In particular, the subaggregate A of all elements of M which precede a definite element m is called a *segment* of M— or the segment determined by M.

E.g., In $\{\ldots -5, -3, -1, 1, 3, 5, \ldots, \ldots, -4, -2, 2, 4, \ldots\}$ consider the aggregate A of all odd numbers. It is an initial part since for any element in A, all preceding numbers are odd and therefore belong to A. But it is not a segment since there is no number in the given aggregate such that all numbers preceding it are only odd numbers.

Definition 14: If the well ordered aggregate M is similar to a segment of the well ordered aggregate N, then the ordinal number μ of M is said to be *less* than that of $N(= v)$ or conversely v is *greater* than m. (We write $\mu < v$.)

$$\mu < \mu+1 < \mu+1+1 < \ldots$$

(It is necessary to prove that of two different ordinal numbers one is always less than the other.)

Also, ω is now seen to be the least transfinite ordinal number. The array of ordinals is now expanded to infinity.

$0, 1, 2, \ldots \omega, \omega+1, \ldots \omega^2, \ldots \omega^\omega, \omega^\omega+1, \ldots\ \omega^{\omega^\omega} \ldots\ \omega^{\omega^{\omega^{\omega \cdots}}} = E,$ $E+1, \ldots.$

Addition and multiplication can now easily be defined as above.

So much for the expansion of the number system into the transfinite and the development of the theory of aggregates. Innumerable problems and the development of the theory have been omitted here. The

[9] Hobson calls them normally ordered aggregates. Vol. 1, pp. 211–12.

interested reader can consult the literature cited.[10] Axiom sets have been given for the theory of aggregates but any further consideration is beyond the scope of this work.

It has been maintained that actually we are not dealing with transfinite aggregates but with a law of construction. Thus all we mean by the aggregate of rational numbers is the law of constructing rational numbers.[11] No true infinite exists but only a potentially "becoming" aggregate. It has been argued that merely because we have shown how to set up a bi-univocal correlation between any arbitrary natural number and a unique even number, it does not follow that "therefore" the aggregate of all natural numbers "is bi-univocally imaged on the aggregate of all even numbers."[12] Actually the difficulty seems to lie in the possibility of interpreting the term "all" when we speak of "all elements of a transfinite aggregate." This problem again raises the difficulty of determining what defines the aggregate. If it is necessary to define the aggregate in terms of its elements, then the "all" becomes of outstanding importance (as we shall see in the next chapter, a paradox can be constructed based on this problem). In order to be able to select "any arbitrary element" of an aggregate we must *know* the aggregate in some way. How do we *know* transfinite aggregates which cannot be laid out before us? These may appear to be extra-mathematical questions, but if the concept of aggregate leads to difficulties and contradictions, then what about the structure built upon it?

B. THE ANTINOMIES AT THE FOUNDATIONS OF MATHEMATICS

The entire structure of mathematics can be obtained from the natural number series. The definition of number, as we have seen, rests upon the notion of class, or aggregate. Furthermore, since geometry has been reduced to arithmetic also, and the theory of integrals has been based upon sets of points (i.e., aggregates), the entire structure of mathematics seems to be based upon the notion of aggregate. At this point one of the most interesting illustrations of critical philosophy appears. The analysis of the concept of aggregate leads to extremely interesting results and a fruitful discussion of the nature of the very

[10] Fraenkel's bibliography is a very complete one and should be consulted. Fraenkel's axiom system is itself an important one—*op. cit.*, pp. 268 ff., as also "Axiomatische Theorie der geordneten Mengen," *Jour. of Math.* 155, 1926, pp. 129–158.

[11] Cf. Felix Kaufmann, *Das Unendliche in der Mathematik und Seine Ausschaltung*, pp. 135 ff.; also Brouwer to be discussed later.

[12] *Ibid.*, p. 141. Cf. also Russell's remark *Principles of Mathematics*, 2nd ed., pp. vi-vii and also Lallemand, *Le Transfini*, Paris, 1934, for a non-finitist, non-Cantorian point of view.

foundation of mathematics. This happens because the concept of aggregate shows itself to be a contradictory concept. It leads to amazing antinomies.[13] The discovery of these antinomies (which we shall state in this chapter) not only gave an impetus to foundational investigations but also to logistic investigations.[14]

The first paradox was published in 1897 by the Italian Burali-Forti. In connection with Frege's definition of number, Russell discovered another paradox—which caused Frege to write in the appendix to his work that one of the foundations of his theory was destroyed. This appeared in 1903. In 1905 the Richard Paradox was published. Other paradoxes have been discovered.

It must be emphasized that these are really antinomies—and not merely apparent antinomies due to habits of thought. We saw earlier that for transfinite aggregates the whole was not "greater" than its part. This was not a paradox but seemed strange because we are accustomed to think of finite aggregates. In the paradoxes which follow it will be demonstrated that an assumption "p" leads to the assertion of its contradictory "p is false," while the assumption "p is false" leads to the assertion "p."

BURALI-FORTI'S PARADOX

It is possible to demonstrate in Cantor's theory of aggregates:

(1) Every well ordered series has an ordinal number.

(2) The ordinal number of a series of ordinals up to and including ω, is $\omega+1$.

(3) The series of all ordinal numbers is well ordered and hence has an ordinal number.

Let W be the aggregate of *all* ordinal numbers arranged according to magnitude. Then by (3) W has an ordinal number—call it ω_1. But by (2), ω_1 is greater than every ordinal in W. However, since W is the aggregate of *all* ordinals, it contains ω_1. Therefore, ω_1 is less than the ordinal of W which is ω_1.

The same thing is true of the aggregate K of all cardinal numbers.[15] We can show that there is a greater cardinal to any cardinal. It can

[13] The literature on the subject has become great, although of recent years interest seems to have died down. The best source for the beginner is in J. Jörgenson, *Treatise of Formal Logic V*, III, pp. 37 ff., and Chap. 12. Cf. also A. Fraenkel, *Einleitung*, pp. 209 ff.; Max Black, *The Nature of Mathematics*, pp. 97 ff.; F. Kaufmann, *Das Unendliche in der Mathematik*, pp. 190 ff.; R. Carnap, *The Logical Syntax of Language*, pp. 211 ff.; Hobson, *op. cit.*, pp. 244 ff.; and F. P. Ramsey, *Foundations of Mathematics*, London, 1931.

[14] Cf. Basilio Mania, "L'infini mathematique et L'évolution de la logique," *Actes du Congrès de Phil. Scientifique*, Paris, 1936, VI, pp. 51 ff.

[15] Cf. Fraenkel, *op. cit.*, p. 212.

also be demonstrated that the sum of all cardinal numbers of an aggregate of cardinal numbers is a cardinal greater than any in the aggregate.[16] Consider the aggregate K of *all* cardinal numbers, and take the sum Σ of all cardinals in K, which is a cardinal greater than any in K, yet is itself in K and Σ is therefore greater than itself.

In these two paradoxes we find the concept of *aggregate*, of *all*, of *ordinal* number, and of *cardinal* number used. The aggregates set up are clearly transfinite ones. These paradoxes involve the use of concepts belonging to mathematics, as well as logic.

THE RICHARD PARADOX

Consider all possible combinations of the letters of the alphabet. Consider that subaggregate of these combinations which define all decimals definable in a finite number of words. This is a countable aggregate and can therefore be counted. Using the diagonal method we can define a decimal N not in the counting. If in the counting the nth digit in the nth decimal is p, then in N put for that digit $p+1$ (if $p = 9$ put 0). N is thus different from every decimal in the counting at least in the nth digit. But the definition of N just given contains a finite number of words. It should therefore appear in the counting.[17]

This paradox involves the whole concept of *definability* as well as Cantor's *diagonal method.*[18]

Other illustrations of paradoxes based on the concept of definability have been given, but for our purposes this is sufficient.

THE RUSSELL PARADOX

This is probably the best known of all the paradoxes.

All classes may be divided into classes which contain themselves as elements and classes which do not contain themselves as elements. For example, the class of all abstractions is itself an abstraction and therefore is a member of itself; but the class of apples is not itself an apple and is therefore not an element of itself. Consider the class M of all classes which *do not* contain themselves as elements, i.e., M contains as its elements classes like the class of apples, of women, etc. We raise the question as to the type of M itself. Is M a member of itself or not?

Let us argue by the indirect method. Suppose M is a member of itself. Then M satisfies the property common to all elements of M,

---

[16] *Ibid.*, pp. 67 and 95.

[17] For the general logical problem involved in this paradox the reader should consult A. Church, "The Richard Paradox," *American Mathematical Monthly*, XLI, June–July, 1934, pp. 356–61.

[18] Fraenkel feels Richard's diagonal method is different in essence from Cantor's (p. 217) but the difference is not of much importance. *See* O. Hölder, *Die Mathematische Methode*, p. 549.

i.e., they are not members of themselves. Therefore, M is not a member of itself. This contradicts our hypothesis and therefore shows it to be false. Assume then that M is not a member of itself. If M is not a member of itself then it satisfies the property common to all elements of M and it therefore is a member of itself. Each assumption *M is a member of itself* and *M is not a member of itself* implies the other, i.e., "p implies p is false."[19]

Involved in this paradox seem to be the concept of *class*, the concept of *element* of a class, the use of *indirect methods of proof*, and the use of the fact that *if a proposition implies its own falsity, it is therefore false.*

PARADOX OF PREDICABILITY

The Russell paradox seems to be one of a type. Another example of what appears to be the same general type of paradox is now given.

All terms which can be predicated of themselves will be considered as the elements of the *class of predicable terms*—P. All terms not predicable of themselves will be the elements of the *class of impredicable terms*—P'. Of course every term is either predicable or not predicable of itself, i.e., is either an element of P or of P'. Furthermore, no term can be both predicable and not predicable of itself—i.e., no term is an element of both P and P'. Thus *abstract* is a predicable term, i.e., abstract is an element of P, while *sweet* is an impredicable term. Now we ask: Is *impredicable* an element of P or of P'? Suppose it is an element of P, i.e., it is a predicable term. Then we have the proposition "impredicable is impredicable." But we have assumed it to be predicable—therefore our assumption that it is predicable leads to the assertion that it is impredicable. Again, the assumption that impredicable is impredicable really means it is predicable. Therefore we have a contradictory class, i.e., the class of impredicable terms is a paradoxical class.

Note here that as in Russell's paradox, the paradox results when we define a class by means of a *negative* property. Russell's paradox sets up the class of all classes *not* containing themselves as elements. The above paradox sets up the class of *non*-predicable (= impredicable) concepts. The significance of this (if there is any) seems to have been overlooked. Why should paradoxical classes seem to prefer *negative concepts*? The same is true of Berry's Paradox.

BERRY'S PARADOX

We can divide all whole numbers into two classes. One class will contain all numbers definable in words containing fewer than nineteen

[19] For an attempt to explain this based on a distinction which is confused, cf. McTaggert, "Propositions Applicable to Themselves," *Mind*, XXXII, 1923, pp. 462–64, Reprinted as Chap. 8 in McTaggert, *Philosophical Studies*, New York, 1934.

syllables, and the other class all numbers definable in words containing nineteen or more syllables. Now consider the smallest number that can be expressed in words in not fewer than nineteen syllables: this number is 111,777 — one| hun|dred| and| e|lev|en| thou|sand| se|ven| hun|dred| and| sev|en|ty| se|ven. But the expression: "the| least| in|te|ger| not| nam|able| in| few|er| than| nine|teen| syl|la|bles|"—names it in 18 syllables. Therefore, the number not namable in fewer than nineteen syllables is namable in fewer than nineteen syllables—a contradiction.[20]

These paradoxes are seen to be of two kinds: 1. Those which use mathematical concepts; 2. those which use purely logical concepts. Fraenkel calls these "logical" and "epistemological" paradoxes[21]. However, they all have in common the fact "that something is asserted about some collection of objects whereby apparently further objects are introduced belonging and at the same time not belonging to the collection in question."[22] If we consider an aggregate to be defined by a propositional function $f(x)$, then the paradoxes seem to arise because in making substitutions for x we can get $f(x_1)$ and $\sim f(x_1)$, i.e., x_1 is shown to satisfy and not to satisfy $f(x)$. Such a state of affairs clearly indicates that something is wrong.

Since the discovery of these paradoxes a great deal of literature on the subject has arisen and many attempts at a solution have been offered. More and more, however, the direct attack on the paradoxes was replaced by a thorough investigation of the foundations of logic and mathematics. The fact that fundamentally we are concerned with a problem belonging to logic necessitated an investigation of logic. This was especially true because actually paradoxes were not new in logical theory. (The Epimenides paradox is probably the best known one. Epimenides was a Cretan who one day, to plague logicians for many centuries, said, "All Cretans are liars," and then wanted to know whether what he said was true or false.[23] If true, then Epimenides is telling a lie. But if his statement is a lie, it is false. If false, then some Cretans are not liars and Epimenides might be one of them and therefore is telling the truth.)

So far as mathematicians are concerned, there appears to be an easy way out. That is to insert in mathematics an axiom which excludes such a paradoxical aggregate. This *avoids* the paradoxes but does not solve them. Furthermore, such a solution in no way eradi-

[20] The investigation of those paradoxes has led to extremely interesting and important results (cf. Carnap, *op. cit.*, pp. 211–20) particularly with respect to such antinomies as Russell's which are linguistic in type.

[21] Fraenkel, *op. cit.*, pp. 210 and 214.

[22] Jörgenson, Vol. 3, *op. cit.*, pp. 166–67.

[23] Cf. the recent discussion of this paradox by Alexandre Koyré, "The Liar," *Philosophy and Phenomenological Research*, VI, No. 3, March, 1946, pp. 344 ff.

cates the root of the occurrence of the paradoxes. (This method is used by Schoenflies and Fraenkel.) For the logistic approach, which reduces mathematics to logic, there is needed either some explanation or prohibition of types of predicates or a new definition of class. Russell's theory of types is an attempt to rid logic of these illegitimate totalities.

Another method of solution lies in the attempt to set up a definition of class (or aggregate) which will not lead to contradictions. This becomes the problem of consistency to which the Formalists direct their attention.

Since some aggregates which lead to paradoxes are infinite aggregates, it is possible to avoid them by denying the existence of such aggregates,[24] or by denying the applicability of the method of proof by means of which the paradoxes are deduced, to infinite aggregates. Such is the position of the Intuitionists.

Weyl has said that mathematics is the science of the infinite.[25] It has been felt that the theory of the infinite has led to a precision in the problem of the nature of the infinite. But the precision seems to have generated problems in the very roots of logic.

We shall now turn to a consideration of the various theories of the foundations of mathematics.

---

[24] Cf. F. Bernstein, "Die Mengenlehre Georg Cantors und der Finitismus," *Jahresbericht der Deutschen Mathematiker-Vereinigung*, Vol. 28, 1919, pp. 63–78.

[25] H. Weyl, *Open World*, Yale Press, 1931, p. 7.

CHAPTER EIGHT

THE FOUNDATIONS OF MATHEMATICS

I. LOGISTIC FOUNDATIONS[1]

We have already studied the logistic derivation of number from logic. Hence we know that the fundamental position of logistics, so far as the foundations of mathematics are concerned, is that all concepts of mathematics can be derived from the concepts of logic by means of nominal definition.[2] The study of the logistic foundations of mathematics resolves itself then into the construction of a logic which will yield all of mathematics and none of the paradoxes of mathematics, as well as the problems involved in logical theory proper. We shall here briefly outline the logical system of Russell-Whitehead, indicating where new axioms are needed to construct mathematics. A great deal of this will repeat what was said in the chapter on the logistic definition of number, but will now be said more systematically and in some detail. We shall follow in our discussion the little book of D. Hilbert and his pupil W. Ackermann, *Grundzüge der Theoretischen Logik* (Berlin, 1928).[3] Hilbert is the leader of the formalist school of thought. His close affinity to logistics will be evident from our use of his book.

Logic studies the relations between propositions, independent of what each proposition is about. In logic we want to know what we can say about certain propositions when others we have given are known to be true or false. Thus from "all *a* is *b*" taken as true, can we always say "some *a* is *b*" is also true? From "all *a* is *b*" and "all *b* is *c*" follows the truth of "all *a* is *c*"—no matter what *a*, *b*, and *c* stand for. Thus from "all men are lions" and "all lions are cats" *follows* "all men are cats"—even though the statements are not valid of actual men, lions, and cats. Truth in logic merely means "follows from" or "can

---

[1] For a good discussion cf. M. Black, *Nature of Mathematics*, 1933, pp. 15–144; J. Jörgenson, *A Treatise of Formal Logic*, 1931, Vol. III; R. Carnap, *Die logizistische Grundlegung der Mathematik*, Erkenntnis, Vol. 2, 1931, pp. 91–105; R. Carnap, *Abriss der Logistik*, 1929; B. Russell, *Principles of Mathematics*, and *Introduction to the Philosophy of Mathematics*.

[2] The reader desiring a more detailed introduction to symbolic logic will find the recent text: John Cooley, *A Primer of Formal Logic*, 1942, excellent. Also cf. S. Langer, *Introduction to Symbolic Logic*, 1937; C. Lewis and C. Langford, *Symbolic Logic*, 1932.

[3] The *Journal of Symbolic Logic* contains a complete bibliography of this field.

be deduced from." In general, logic studies the "conditions under which the truth of a proposition p can be deduced from the truth of a set of *propositions* $P_1, P_2 \ldots P_n$." Logic, from this point of view, is the basis of all other sciences. By a proposition we mean any sentence of which it is meaningful to say that it is true or false, e.g., "mathematics is a science," "9 is a prime number." It is not meaningful to say of a sentence "How beautiful!" that it is true or false. Logic does not concern itself, therefore, with such sentences, although "syntax" may. (We shall see later that the metalogic of Hilbert and the work of Chwistek are in the field of the syntax of logic.)

LOGIC OF PROPOSITIONS

The relations of propositions are expressed in verbal form by "and," "or," "not," "if-then." We shall let $p, q, r \ldots$ stand for propositions. We write

$\sim p$ for *not-p*, i.e., if p is true, $\sim p$ means p is false.
  $\sim p$ denotes the contradictory to p.

$p \& q$ for *p and q*, i.e., p and q are true together. (This is the dot of the Russell chapter.)

$p \vee q$ for *p or q*, i.e., either p or q is true—possibly both.

$p \rightarrow q$ for *if p then q*, i.e., either p is false of q is true. (The only case in which $p \rightarrow q$ is false is that in which p is true and q is false. The reader should make this very clear to himself. This is the $\supset$ of the Russell chapter.)

From this definition of $\rightarrow$ it follows that if p is given as true, then q is true.

$p \equiv q$ for *p and q are equivalent*. It is defined as $p \rightarrow q$ and $q \rightarrow p$. In our symbolism

$$(p \equiv q) =_{Df.} (p \rightarrow q) \& (q \rightarrow p).$$

This really means that p and q are both true or both false, e.g., if p is "$2 > 3$" and q is "black is white," then $p \equiv q$.

Consider the possible combinations in p and q

p	q	$p \rightarrow q$
T	T	T
T	F	F
F	T	T
F	F	T

We explain the second and third cases: $p \rightarrow q$ means that either p is false or q is true. Now if p is true and q is false, then neither case occurs and $p \rightarrow q$ is false.

If p is false and q is true, then both cases are satisfied and $p \rightarrow q$ is true.

(The array of combinations of T and F is called a matrix. The student should familiarize himself with it.)

In many cases we have two symbols for the same thing, e.g., $\sim(\sim p)$, and p. We shall denote their equivalence by an $=$ sign. Note that this is not a definitional sign but merely denotes their equivalence. Thus we write:

$\sim(\sim p) = p$

$p \& q = q \& p$

$p \& (q \& r) = (p \& q) \& r$

$p \vee q = q \vee p$

$p \vee (q \vee r) = (p \vee q) \vee r$

$p \& (q \vee r) = (p \& q) \vee (p \& r)$

$p \vee (q \& r) = (p \vee q) \& (p \vee r)$

We shall set up a matrix for one of these to show the equivalence of $p \& (q \& r)$ with $(p \& q) \& r$.

1	2	3	4	5	6	7
p	q	r	$p \& q$	$p \& r$	$p \& (q \& r)$	$(p \& q) \& r$
T	T	T	T	T	T	T
T	T	F	T	F	F	F
T	F	T	F	F	F	F
T	F	F	F	F	F	F
F	T	T	F	T	F	F
F	T	F	F	F	F	F
F	F	T	F	F	F	F
F	F	F	F	F	F	F

Notice that every notation in Column 6 is the same as that in Column 7. Therefore 6 and 7 are equivalent.

Columns 1, 2, and 3 give all possible combinations of three true and false propositions. The first three columns give us the truth or falsity of the constituents of the given formula. Columns 4 and 5 build up the more complex elements, while Column 6 is the truth or falsity of the left-hand side and 7 that of the right-hand side. Let us explain the combination F, T, F. Column 4 has $p\&q$. & means both p and q are true together. But p is F. Therefore $p\&q$ must also be F. Column 5, $q\&r$ likewise gives F. Column 6, $p\&(q\&r)$ now is F & F. But again by definition of & this is F. Column 7 $(p\&q)\&r$ also gives F.

We also need:

a) $\sim(p\&q) = \sim p \vee \sim q$

b) $\sim(p \vee q) = \sim p \& \sim q$

c) $p \rightarrow q = \sim(p\&\sim q)$

We can prove $p \rightarrow q = \sim p \vee q$ as follows:

given $p \rightarrow q = \sim(p\&\sim q)$

substituting in (a), we get $\sim(p\&\sim q) = \sim p \vee \sim(\sim q)$

By $\sim(\sim p) = p$, we get $\sim p \vee q$.

$\therefore p \rightarrow q = \sim(p\&\sim q) = \sim p \vee q$.

We can further prove $p \rightarrow q. = \sim q \rightarrow \sim p$ as follows:

$p \rightarrow q = \sim p \vee q$

$\sim q \rightarrow \sim p$, by substitution ($p = \sim q$; $q = \sim p$)

gives $\sim(\sim q) \vee \sim p = q \vee \sim p$, which is $\sim p \vee q$.

The reader can prove:

$p \vee q = \sim(\sim p \& \sim q)$.

The truth or falsity of any combination of propositions constructed out of the basic propositions $p_1 p_2 \ldots p_n$ using the logical symbols &, $\vee$, $\rightarrow$, $\equiv$, $\sim$, depends only on the truth or falsity of the constituent propositions.

The first problem of logic is then to discover those combinations of propositions which are always true, i.e., true no matter whether the constituent propositions are true or false. Such combinations are called *tautologies* [e.g., $(p\&\{p \rightarrow q\}) \rightarrow q$]. In order to solve this problem we set up four fundamental propositions and two rules of procedure. (Note that these rules of procedure cannot be guaranteed within the given system. They are really extra-logical.)

FUNDAMENTAL PROPOSITIONS

a) $p \vee p \rightarrow p$
b) $p \rightarrow p \vee q$
c) $p \vee q \rightarrow q \vee p$
d) $(p \rightarrow q) \rightarrow [(r \vee p) \rightarrow (r \vee q)]$

(We leave their interpretation to the reader.)

RULES OF PROCEDURE

α) Rule of substitution:

If we have a tautology and replace any given symbol p everywhere it occurs by any other symbol or combinations of symbols, another tautology will result.[4] E.g., given $p \vee \sim q$. Replace p by $q \vee \sim p$. We then get $(q \vee \sim q) \vee \sim(q \vee \sim q)$. (This is an inexact expression of the rule but is sufficient for our purposes. Cf. Hilbert-Bernays, *Grundlagen der Math.*, Vol. I, Berlin, 1934, pp. 48–50.)

β) The syllogistic principle:

If A and $A \rightarrow B$ are tautologies then B is also.

(Note that we can obtain propositions with & by using $\sim(p \vee q) =_{Df.} (\sim p \& \sim q)$. The sign $\rightarrow$ is only an abbreviation $p \rightarrow q =_{Df.} \sim p \vee q$. Thus $p \vee p \rightarrow p$ really means $\sim(p \vee p) \vee p$—which is really $p \vee \sim p$—the general form of a tautology.)

We shall show how one derives other tautologies from these by proving:

$$p \vee (q \vee r) \rightarrow q \vee (p \vee r).$$

This was originally taken as an additional fundamental proposition, but P. Bernays showed it could be derived from the others.[5]

(The reader should demonstrate the tautological character of the four fundamental propositions by the matrix method involving the meaning of $\vee$ and &.)

We shall need for our proof two auxiliary propositions:

(1) $(p \rightarrow q) \rightarrow [(r \rightarrow p) \rightarrow (r \rightarrow q)]$.

Proof:

By fundamental proposition (d) we have:

$(p \rightarrow q) \rightarrow [(r \vee p) \rightarrow (r \vee q)]$

[4] Cf. Hilbert and Ackermann, *Grundzüge der Theoretischen Logik*, p. 23, and Black, *op. cit.*, p. 46. We word this differently than either source.

[5] Bernays, "Axiomatische Untersuchungen des Aussagenkalküls der P.M.," *Mathematische Zeitschrift*, 25, 1926.

In this let $r = \sim r$ and we have

$(p \rightarrow q) \rightarrow [(\sim r \vee p) \rightarrow (\sim r \vee q)]$

but $\sim r \vee p =_{Df.} r \rightarrow p$

$\sim r \vee q =_{Df.} r \rightarrow q$

Substituting, we get (1).

(2) If $(p \rightarrow q)$ and $(q \rightarrow r)$ are tautologies, then $p \rightarrow r$ is also.

Proof:

In (1) replace p by r, and r by p. Do not alter q. Then we have

$(r \rightarrow q) \rightarrow [(p \rightarrow r) \rightarrow (p \rightarrow q)]$

If now $r \rightarrow q$ is a tautology and

$(r \rightarrow q) \rightarrow [(p \rightarrow r) \rightarrow (p \rightarrow q)]$, then

$[(p \rightarrow r) \rightarrow (p \rightarrow q)]$ is,

and if $p \rightarrow r$ is a tautology and $(p \rightarrow r) \rightarrow (p \rightarrow q)$ is, then $p \rightarrow q$ is. Thus if $p \rightarrow r$ and $r \rightarrow q$ are tautologies, $p \rightarrow q$ is also.

Proof of $p \vee (q \vee r) \rightarrow q \vee (p \vee r)$:

$r \rightarrow r \vee p \rightarrow p \vee r$ by fundamental proposition (b) and (c); therefore $r \rightarrow p \vee r$ by (2).

$[r \rightarrow (p \vee r)] \rightarrow [(q \vee r) \rightarrow q \vee (p \vee r)]$ by (d).

Also since $q \vee r \rightarrow q \vee (p \vee r)$

*we obtain $p \vee (q \vee r) \rightarrow [p \vee \{q \vee (p \vee r)\}]$ by (d).

We start again:

$p \rightarrow p \vee r \rightarrow r \vee p$

therefore $p \rightarrow r \vee p$.

Let p be replaced by $p \vee r$, and r by q

We then have $(p \vee r) \rightarrow q \vee (p \vee r)$.

Now since $p \rightarrow p \vee r \rightarrow q \vee (p \vee r)$

then $p \rightarrow q \vee (p \vee r)$.

Adding $q \vee (p \vee r)$ to both sides, by (d) we get

$[q \vee (p \vee r)] \vee p \rightarrow [q \vee (p \vee r)] \vee [q \vee (p \vee r)]$

but by (a), $[q \vee (p \vee r)] \vee [q \vee (p \vee r)] \rightarrow q \vee (p \vee r)$

therefore

**$[q \vee (p \vee r)] \vee p \rightarrow q \vee (p \vee r)$.

Consider together * and **, we have

$p \vee (q \vee r) \rightarrow p \vee [q \vee (p \vee r)] \rightarrow [q \vee (p \vee r)] \vee p \rightarrow q \vee (p \vee r)$

therefore $p \vee (q \vee r) \rightarrow q \vee (p \vee r)$

LOGIC OF PROPOSITIONAL FUNCTIONS[6]

So far we have been concerned with the relations between actually true or false propositions. However, we frequently reason as follows: If x is a prime number other than 2, then x is not divisible by two. In our symbolism, we have $\sim$(x is a prime number) $\vee$ (x is not divisible by two). Neither expression in parentheses can be called a proposition yet the relation seems to be a necessary one. Actually, of course, the proposition is usually a special case of some such general form. A form which contains an undetermined constituent, but is capable of becoming a proposition when the undetermined constituent is filled in, is called a *propositional function*. As in mathematics, the undetermined constituent is called a *variable*. We shall denote a propositional function by $P(x)$ or $\phi(x)$ or some such form. Thus if we consider "x is a prime number," we let P = "is a prime number," $P(x)$ then means "x is a prime number." If further ϕ means "is not divisible by two," then our illustration becomes $P(x) \rightarrow \phi(x)$. We can make this a proposition by letting $x = 5$; then we have $P(5) \rightarrow \phi(5)$, i.e., "5 is a prime" implies "5 is not divisible by two." The relations here become those of the logic of propositions.

Given a function $P(x)$. If x is replaced by a definite value, say x_1, then $P(x_1)$ is a proposition which is either true or false. If $P(x_1)$ is true, then we say that the *truth-value* of $P(x)$ for x_1 is *truth*, otherwise *falsity*. Thus a propositional function has two possible truth values, truth or falsity. That is, when a function becomes a proposition, the resulting proposition is either true or false. On this basis if a function $P(x)$ is not meaningless for $x = x_1$ and if we can show $P(x_1)$ cannot have the truth-value *truth*, then $P(x_1)$ must have the truth-value *falsity*. (This is the *reductio ad absurdum* proof which will form the target of Brouwer's attack.)

Suppose P means "is less than," then $P(2,3)$ is "2 is less than 3," which may also be written $<(2,3)$. Clearly $<(x,y)$ is not true for all values of x and y. But there are x's and y's (e.g., $x = 5$, $y = 2$) such that the truth value of $<(x,y)$ is falsity. On the other hand, if we restrict the range of our variable to human beings and let ϕ mean "is a mortal," then $\phi(x)$ always has the truth-value, truth. These two ideas require symbolic expression. *All* is symbolized by (x) and *there is* by (Ex). Thus $(x)P(x)$ means "$P(x)$ is true for all X," whereas $(Ex)\sim P(x)$ means "there is an x such that $P(x)$ is false." E.g., if L means "is a point through which a line may be passed," then $L(x)$ means "x is a point through which a line may be passed" and $(x)L(x)$ means "for all points x, x is a point through which a line may be passed." Also

[6] We omit a special discussion of logic of classes here because a class is defined by means of a propositional function.

$(Ex)L(x)$ means "there is a point x, such that x is a point through which a line may be passed." (x) and (Ex) are called *quantifiers*—(x) is a general quantifier; (Ex) is an existential quantifier. We must distinguish carefully between, e.g., $\sim(x)\phi(x)$, and $(x)\sim\phi(x)$. We can have more general expressions, e.g., $(x)(y)P(x,y)$ read "for all x and for all y, $P(x,y)$ is true," or $(x)(Ey)P(x,y)$, read "for all x there is a y such that $P(x,y)$ is true," etc. Two other illustrations: $(x)[\phi(x) \rightarrow \psi(x)]$, read "for all x, 'if $\phi(x)$ then $\psi(x)$'," i.e., the quantifier pertains to the whole expression in the brackets. $(Ey)(x)A(x,y) \rightarrow (x)(Ey)A(x,y)$, read "*if* there is a y such that for all x, $A(x,y)$ is true, *then* for all x there is a y such that $A(x,y)$ is true." *The two expressions are not identical.*

As an illustration of the translation of sentences into symbolic form Hilbert and Ackermann (pp. 48–49) give:

"*1. Every number has one and only one immediate successor.*

2. There is no number such that 1 is its immediate successor.

3. Every number different from 1 has one and only one immediate predecessor.

In symbolic form:

Let $F(x,y)$ be 'y is the immediate successor of x,'
$= (x,y)$ be 'x equals y.'

1′. $(x)(Ey)\{F(x,y)\&(z)F(x,z) \rightarrow =(y,z)\}$ read 'for all x, there is a y such that if y is immediate successor of x and if for all z, z is the immediate successor of x, then y equals z.'

2′. $\sim(Ex)F(x,1)$

3′. $(x)\{\sim=(x,1) \rightarrow (Ey)[F(y,x)\&(z)(F(zx) \rightarrow =(y,z))]\}$"

AXIOMS OF THE LOGIC OF PROPOSITIONAL FUNCTIONS

a) $p \vee p \rightarrow p$

b) $p \rightarrow p \vee q$

c) $p \vee q \rightarrow q \vee p$

d) $(p \rightarrow q) \rightarrow (r \vee p \rightarrow r \vee q)$

e) $(x)F(x) \rightarrow F(y)$

f) $F(y) \rightarrow (Ex)F(x)$.

The procedure from this point involves nothing especially new so far as derivation is conceived.

When discussing the propositional function we note that it can be considered in two ways: (1) A propositional function $P(x)$ states a property P of x. When the P is considered we speak of the *intension* of the function. (2) A propositional function $P(x)$ denotes a set of values which can be assigned to x. When $P(x)$ is considered from the point of view of the range of its argument we speak of its *extension*.

Thus if P = "is a rational animal," then the intension of $P(x)$ is rational animal; the extension, the objects which are rational animals.

Now if $P(x) \rightarrow R(x)$ for all values of x, then we say $P(x)$ *formally implies* $R(x)$. If two functions formally imply each other they are said to be equivalent. (This really means in the language of the theory of aggregates that the bi-univocal image between the extension of $P(x)$ and $R(x)$ is one of identity.) In symbols:

$$\phi(x) \equiv \psi(x) =_{Df.} (x)\{[\phi(x) \rightarrow \psi(x)][\psi(x) \rightarrow \phi(x)]\}.$$

EXTENDED FUNCTIONAL CALCULUS

So far we have been concerned with functions whose arguments were individual objects. It is of course possible to have functions whose arguments are also functions. For example if R = "is a reflexive relation," then $R(F)$ means "The function F is a reflexive relation"—where F may be the function "is equal to." Another illustration is the one of equivalence between propositional functions. We write "Equ. (P,Q)" to mean "$P(x)$ is equivalent to $Q(x)$." Other relations may be defined: "In (P,Q)" (read: "$P(x)$ is incompatible with $Q(x)$"). This is also $(x)[\sim P(x) \vee \sim Q(x)]$. "Imp. (P,Q)" (read "$P(x)$ implies $Q(x)$.") This is also $(x)[P(x) \rightarrow Q(x)]$.

We can add quantifiers "$(P)(Q)$Eq.(P,Q)," read "P is equivalent to Q for all P and all Q." With this notion we can define other functions; e.g., identity (written $\equiv(x,y)$) may be defined as: $\equiv(x,y) =_{Df.} (F)[F(x) \rightarrow F(y)][F(y) \rightarrow F(x)]$.[7]

The extended calculus makes possible the construction of the concept of number. As we saw before, quantity is a property. Since each object is *one*, the quantity of a group of objects cannot be predicated of the objects in the group. The number is a property of that concept under which the counted objects are subsumed. This means that number is a predicate of a function. We repeat some of the definitions given before in a somewhat different form:

$0(F) : \sim(Ex)F(x)$, i.e., 0 is predicated of F if no object exists which satisfies F.

$1(F) : (Ex)[F(x) \& (y)\{F(y) \rightarrow \equiv(x,y)\}]$

$2(F) : (Ex)(Ey)\{\sim\equiv(x,y) \& F(x) \& F(y) \& (z)[F(z) \rightarrow \equiv(x,z) \vee \equiv(y,z)]\}$

Other concepts of mathematics and the concepts of the theory of aggregates as well as the paradoxes can be translated into these symbols.

[7] This involves difficulties; cf. Black, *op. cit.*, p. 71. However, if we treat x and y as symbols of the same thing, the definition is valid provided (F) is not infinite. If x and y are symbols only of themselves, I cannot see that identity means anything except substitutability. But how do we know that different positions are irrelevant?

THE THEORY OF TYPES AS A SOLUTION TO THE PARADOXES[8]

Russell's theory of types is based on the view, arrived at by logical analysis of the paradoxes, that they all originate from a vicious circle, due to supposing that a collection of objects may contain members which can only be defined by means of the collection as a whole. Such collections are said not to exist and propositions regarding such are said to be meaningless.[9]

The theory of types breaks up into a number of parts, only one of which we shall discuss here. In logic, we have classes, propositions, properties, and relations. We can have classes of individuals and of classes; propositions about individuals and about propositions; properties of individuals and of properties, and relations between individuals and between relations. We can therefore, set up a theory of types for each of these, i.e., a theory determining the use of reflexivity so far as each of these are concerned.

In the first place it should be noticed that to every aggregate there are correlated propositional functions which define it, and conversely to every propositional function (or set of propositional functions) there is correlated an aggregate. Hence, the whole theory of aggregates reduces to the theory of propositional functions (or conversely the theory of propositional functions reduces to the theory of aggregates).[10]

Secondly, it will be noticed that many of the paradoxes result when we ask whether the function which defines[11] the class can itself satisfy the function, i.e., if $f(x)$ defines the class of all classes not members of themselves, what is the truth value of $f(f(x))$? This really means that some of the paradoxes result in the extended functional calculus. If we are to avoid these paradoxes then some restrictions must be placed upon the range of variability of the argument. This is the heart of Russell's Theory of Types and Orders.

For a propositional function to be well defined we must know beforehand what the possible values for its argument are, so that these possible values (the aggregate of these is called the *range of significance* of the function) must be defined beforehand. This must clearly be done independently of the function itself.

If we consider the argument of a function as different from the function, then we can say that the range of a function cannot include anything which depends on the function itself. For example, if we call adjectives whose meaning can be predicated of themselves "auto-

---

[8] Cf. Russell, *Principles*, p. 523, *Principia Math.* Vol. I; Black, *Nature*, pp. 100 ff.; and Jörgenson, *op. cit.*, Vol. 3, pp. 168 ff. and Vol. 1, pp. 206 ff.

[9] *Ibid.*, p. 43.

[10] Tarski, *Einführung*, p. 44.

[11] Definitions in which such a process occurs are called *nonpredicative.* Cf. Fraenkel, *op. cit.*, pp. 247–54. The theory of types may be said to prohibit nonpredicative functions of a definite kind.

logical" (for example, the adjective "short" is "short"), and those which cannot, "heterological" (for example, the adjective "sweet" cannot be predicated of "sweet"), then the contradiction which results when we ask "is heterological, heterological?" arises when we try to include in the range of the definition that which is being defined. This is a special case of the antinomy on "predicable-impredicable" discussed previously. We then are led to construct a *hierarchy of types* which prevents us from constructing paradoxical aggregates—since it prevents us from constructing an aggregate (defined in extension by the function) which can have itself as an element. Propositional functions, for whose argument we substitute the function, do not become propositions, but meaningless symbols. Thus to ask "Is $f(f(x))$ true or false?" is a meaningless question. (This would appear to remain a function since x is still undetermined, but it is actually on the basis of the theory of types meaningless.) This means that only aggregates which are ranges of significance for some function are legitimate.[12]

A range of significance defines the *logical type* of the function. Any range of significance which presupposes another range of significance is said to be of *higher type* than that which it presupposes. Thus every proposition about an aggregate is of higher type than the proposition defining the aggregate itself. Since, also, a class is defined by the totality of its elements, we cannot meaningfully ask whether the class is a member of itself. This is equivalent to attempting a definition of a class in terms of itself—which is a kind of vicious circle. In this way number, as a class of classes, is of higher type than the element classes and these are of higher type than the elements in the element classes.

Briefly, we can state the theory of types again as "no class can contain itself as a member and no propositional function can take itself as argument, or anything which assumes itself." The whole appears to involve a circle in definition and Russell calls this the "vicious circle principle." Since any assertion about *all* classes or *all* propositions would necessarily consider these as aggregates of which they were members (i.e., a proposition about the class of all propositions would be a proposition and therefore be in the class of all propositions), the theory of types would forbid any such assertions. However, this cannot be all of the theory of types since obviously in logic we frequently must make assertions about "all true propositions," etc.

At this point a familiar distinction of classic logic offers us a clue—the distinction between the collective and distributive use of *all*. The collective use of *all* refers to the aggregate as a whole. For example, when we speak of *all* the army—we mean the army as a unit. The distributive use of *all* refers to each individual. For example, when we

---

[12] Cf. B. Russell, "Mathematical Logic as Based on the Theory of Types," *American Journal of Math.* 30, 1908, p. 222.

say all angles of a triangle are less than 180°, we mean *each* one is. Now when we have a function concerning *all* distributively, the function itself is not implied and hence we have a legitimate function. When the function is about all collectively then the function is illegitimate. What then can we say about the types, i.e., the ranges of significance, of propositional functions?

The lowest type is clearly the "individuals." Individuals presuppose no other aggregate or object to define them, but they are presupposed by all other aggregates. The next type is obtained by generalizing and forming aggregates of individuals. Then any aggregate which presupposes individuals would be of type higher than the individuals in it. The next higher type would be aggregates containing aggregates of individuals; and so on. This can be carried over to the propositions defining the aggregates since, as we said above, every aggregate is defined by a proposition.

It must be pointed out before we proceed that not every x we see in an expression denotes a real variable. For example, "x is a mortal" allows us to substitute any value for x and obtain propositions either true or false. But if we write $(x)\phi(x)$, which means "all x such that $\phi(x)$ is true," we are definitely restricted to such x's as are satisfiers of $\phi(x)$. The first is an example of a real variable, the second of an *apparent* variable. As a little consideration will show, wherever you have a quantifier you have an apparent, or bound, variable.

Let $a, b, c, \ldots$ be individuals. Then each of $a, b, c, \ldots$ is type 0.

Any proposition which presupposes the aggregate of $a, b, c, \ldots$ will define the aggregate of these elements. Hence it will be of Type 1. Such propositions will be of four kinds: $\phi a, \phi b, \ldots$; $\phi(x)$; $(Ex)\phi(x)$; $(x)\phi(x)$. Actually, in this case all x's are apparent variable and we can say "any expression concerning an apparent variable is of higher type than that variable." ($\phi a, \phi b, \ldots$; and ϕx are called *elementary* propositions, $(Ex)\phi x$ and $(x)\phi x$ are called generalized propositions.)

The next type is that which presupposes Type 1; i.e., presupposes aggregates of individuals and therefore defines an aggregate of aggregates of individuals. Such propositions will be $\psi(fa)$, $\psi(fx)$, $(fx)\{\psi(fx)\}$, $(Efx)\{\psi(fx)\}$, etc.

Higher and higher types can be constructed.

What has been said applies to propositional functions as well as propositions. In the case of the arguments of functions we can say that only assertions and functions can be substituted for the segment as are of lower type than the function. E.g., if $F(x) \rightarrow g(x)$, and $F(x)$ and $g(x)$ are both of Type 1, then we can substitute for x only elements of Type 0.

The student should now re-think the paradoxes discussed earlier and note the applicability of the theory of types to their solution.

Axiom of Reducibility: The theory of types avoids the paradoxes but it involves other difficulties in the body of mathematics itself. One of these was indicated in our discussion of irrational numbers, and necessitated Townsend's proof of the theorem that every class of classes converged to a real number. This was in connection with the introduction of irrational numbers as classes of rational numbers. This, from the point of view of the theory of types, means that irrationals are of higher types than rationals. Now if we take a class of irrationals, we have a class of higher type than that of rationals, etc. It therefore seems to follow from the theory of types that there are infinitely many types of real numbers. Geometrically this would mean that a line was of higher type than a point and (if we consider a plane figure as a class of lines) a plane would be of higher type than a line, etc. The theorem cited in Townsend seems to indicate that the theory of types is wrong or needs modification. The possibility of demonstrating that the potency of a plane figure is the same as that of a line would again seem to indicate that the theory of types does too much. As a matter of fact, all theorems dependent upon a Dedekind cut would be suspect, and in many cases invalid (e.g., the theorem that any bounded aggregate of numbers has an upper bound).[13]

To obviate these difficulties involved in the theory of types, the axiom of reducibility is introduced. Recalling that by a predicative function we meant a function whose argument values are individuals, i.e., a function of Type 1, then this axiom reads:

"For every function occurring in the hierarchy of types there exists an equivalent predicative function," or "Every propositional function of one or two arguments is formally equivalent to a predicative function of the same arguments."

If we let $P_n(x)$ be a property of type n, then in symbolic form our axiom would affirm $(P_n)(EP_1)(x)[P_n(x) \equiv P_1(x)]$, i.e., for every P of type n there is a P of type 1 such that for all x_1 $P_n(x)$ implies and is implied by $P_1(x)$.[14]

In the *Principia Mathematica* of Russell-Whitehead, the axiom is written as:

$$*12.1 \quad \vdash :(Ef):\phi x \equiv_x f!x P_p$$

$$*12.11 \quad \vdash :(Ef):\phi(x,y) \equiv_x f!(x,y) P_p$$

This principle is called an axiom because it cannot be proved by the means of the system. But what is more important is that it seems to be an *extra*-logical principle. It is not a tautology—which is a charac-

---

[13] Black, *op. cit.*, p. 107, and Stebbing, *Modern Introduction to Logic*, London, 1933, p. 462, note 1.

[14] Hilbert und Ackermann, *op. cit.*, p. 107.

teristic of logical principles. This means that wherever in mathematics the principle must be used we have a place where a mathematical principle is not derivable from logic alone.[15]

This principle is not, as it appears to be in *Principia*, really part of the deductive scheme of logistics. It is rather a rule for the manipulation of symbols. This is evidenced by the fact that actually one can eliminate the theory of types by using a sufficiently complex system of symbols (cf. Chwistek, Church, Quine). But any statements about the system of symbols—for example, "the system of symbols is complete"—would lead to a violation of the principle. It has been suggested that we might use wooden rings to represent propositional functions and have functions of the same type of equal radius. Any function's argument would have a radius small enough to fit the ring tightly into the function-ring. Since no ring could fit into another of equal or smaller radius, the theory of types would be *shown*.[16] This, of course, still has difficulties—the proposition that "no ring could fit into a ring of equal or less radius" is a proposition about *all* rings and hence violates the theory of types.

Both Hilbert and Ackermann's symbolic expressions, as well as *12.1 and *12.11, violate the principle they are trying to express.

The development of the logistic foundations of mathematics has shown the interdependence of logic and mathematics. It seems that it has raised a storm of problems in logic which has given impetus to a great development in the investigation of logical theory. But the necessity for clearing up these difficulties in logic has weakened the thesis that mathematics is completely derived from logic. The reader should see the importance, in this chapter and the next on Formalism, of becoming acquainted with the field of Symbolic Logic. (There are other axioms, the axiom of infinity and axiom of choice or multiplicative axiom—sometimes also called Zermelo axiom[17]—which are non-tautological and are necessary to the foundation of mathematics. We shall not consider them except to remark that they also weaken the thesis of the identity of mathematics and logic.)

The general feeling today concerning the logistic thesis is, to quote Dubislav,—"it (i.e., logistics) has not succeeded in deriving mathematics completely from logic.... But logistics deserves credit especially for having developed the logical calculus into a powerful instrument for scientific theoretical investigations.[18] Again Jörgenson says: "It

---

[15] Hilbert und Ackermann, *op. cit.*, pp. 109 ff., offer examples of use of this axiom in mathematics.

[16] Black, *op. cit.*, p. 114.

[17] Cf. Church, "Alternatives to Zermelo's Axiom," *Trans. American Math. Society*, Vol. 29, 1927, p. 178 ff.

[18] Cf. Dubislav, *Phil. der Mathematik*, p. 43.

(i.e., logistics) enables us, more than any other science, accurately to determine what primitive ideas and primitive propositions are necessary and sufficient for the construction of any existing deductive theory. . . ."[19]

With this result, a number of alternatives are possible: (1) attempt to reconstruct logistics so that the theory of types is unnecessary or modified; (2) reject the thesis that mathematics is derivable from logic and treat mathematical concepts as mere symbols combined and worked by means of arbitrary rules of procedure in accordance with logical principles; (3) reject the thesis and treat mathematics and logic as independent even to the laws of the relationships between their respective elements (although there is actually some overlapping as we shall see). (1) is illustrated in the work of Ramsey and also in the work of Chwistek. (2) is the work of some of the Formalists. (3) is the work of the Intuitionists.

APPENDIX: CHWISTEK'S NOMINALISTIC APPROACH

Chwistek's work has received slight notice except as an attempt to reconstruct the logistic system. Actually, however, an examination of his work shows that his point of view carries the formalistic notion farther to its extreme and involves many radically semiotic techniques.

As a matter of fact, Chwistek's work may be viewed as an intermediary position between the logistic and extreme formalist approaches. Chwistek accepts the view that mathematics and logic are one; both are derivable from a sort of meta-logic. But at the same time he desires to avoid any suspicion of metaphysics and so treats the symbol and that which it symbolizes as identical. That is, the symbol stands merely for itself. In this respect Chwistek is an extreme formalist, but views his problem as a semantical one.

Chwistek's work divides into two parts: (1) a reconstruction of the theory of types, and (2) his own semantical analysis of the fundamental pattern of mathematics and logic. This analysis has been modified during the years. Among the papers in which Chwistek's work has been developed are the following:

"The Theory of Constructive Types," *Annales de la Société Polonaise de Mathématique*, Vol. 2, 1924, p. 9; Vol. 3, 1925, p. 92 (reprinted by University Press at Cracow, 1925—cited as Const. Types).

"Sur les fondements de la logique moderne," *Atti del V Congresso Internazionale di Filosofia, Napoli*, 1924 (to be cited as Log. Mod.).

---

[19] Jörgenson, *op. cit.*, Vol. 3, p. 200. The reader should consult for a criticism of the logistic thesis, J. R. Weinberg, *An Examination of Logical Positivism*, N. Y., 1936, Chap. 2, and Jörgenson, *ibid.*, Chap. 10.

"Neue Grundlagen der Logik und Mathematik," *Mathematische Zeitschrift*, Vol. 30, 1929, p. 704 (to be cited as G. L. M. 1).

"Une méthode métamathématique d'analyse," *Comptes-Rendus du Premier Congrès des Math. des Pays Slaves*, Warsaw, 1929, p. 254 (to be cited as Met. An.).

"Neue Grundlagen der Logik und Mathematik, Zweite Mitteilung," *Math. Zeit.*, Vol. 34, 1932, p. 527 (to be cited as G. L. M. 2).

"Die nominalistische Grundlegung der Mathematik," *Erkenntnis*, Vol. III, 1933, p. 367 (to be cited as N.G.).

"New Foundations of Formal Metamathematics" (with W. Hetper), *Journal of Symbolic Logic*, Vol. 3, 1938, p. 1 (to be cited as F. M.).

The paper Log. Mod. is an abbreviation of Const. Types. G. L. M. 2 is the same as G. L. M. 1 but simplifies the approach. Probably the best place for the beginner to enter Chwistek's ideas—certainly the most available to the average student in America—is by way of F. M.

In Const. Types, Chwistek reports, he attempted to construct a science of deductive systems without using the axiom of reducibility—which is not a rule of the logical calculus but an axiom of existence. Furthermore, since the rules of direction (i.e., directions for the construction of expressions denoting a proposition or a propositional function) are only partially precise, they needed to be made completely so and all of them listed. Such expressions are called "utility."

As base of the rules of substitution, the primitive form (simple theory) of the theory of types was used. The term *constructive theory of types* was used to denote that the theory does not permit a proof of the existence of an object—unless we have an illustration of that object.

Since there is no essential difference between a propositional function of one variable and a class, symbols for propositional functions of one variable are alone used to represent classes. This is really to eliminate the notion of class entirely. Only such functions are used as are extensional. This means that if two functions are equivalent they are identical.

The purpose of what follows is to start with the idea of function and to construct an elaborate symbolism which will make it possible to construct a new symbol for each type. This is done in such a way that the structure of the symbol makes violations of the theory of types impossible.

The pure theory of types can be abbreviated thus:

a, b, c, d are to represent "constant individuals";

x, y, z are to represent "variable individuals";

$f\{a\}$ denotes "a belongs to the class f";

$T_f(x,a)$ is an abbreviation of $f\{a\} \equiv f\{a\} : f\{x\} \equiv f\{x\}$ and may be read "x is determined by a";

$x'[.f\{x'\}:T_g(x',a).]$ denotes the class f of individuals determined by a. x' is a *noted* variable;

$.f\{y\}:T_0(y,a)$ has the same meaning as $x'[.f\{x'\}:T_0(x',z).]\{y\}$
$(x'').f\{x''\}:T_0(x'',a)$ denotes the proposition "all individuals determined by a belong to class f";
x'' is only an apparent variable;
$h\{x'[.f\{x'\}:T_0(x'a)\}.$ denotes the proposition "the class $x'[.f\{x'\}:T_0(x'a)]$ is the element of class h";
$\hat{u}[.h\{\hat{u}\}.T_1(\hat{u},x'[.f\{x'\}:T_0(x'a).])$ denotes the class h of classes determined by $x'[.f\{x'\}:T_0(x'a).]$.

Two expressions are said to denote *classes of the same type* if their variables are determined by the same expression and if they contain the same letters in addition to their noted or apparent variables—being part of the same utility expressions.[20]

The theory described makes impossible the substitution of an expression of type n for the argument in a function of type $n - 1$. Each type has its own kind of symbol. The axiom of reducibility is unnecessary since each successive type is constructed on the basis of the preceding type.

The variable determined by an expression denoting the class f, denotes an arbitrary class of the same type as f. This simplified form of the theory of types received wide acceptance.

On the basis of this theory, Chwistek constructs a theory of the arithmetic of the cardinal and inductive numbers (Const. Types pp. 48–94). However, since this is a modification of the logistic derivation using the theory of types, we feel it beyond our purpose here.[21] We turn our attention to Chwistek's semantic ideas.

That Chwistek is a member of the logistic school is evidenced by his adherence to the logistic thesis of the derivability of mathematics from logic.[22] But he desires to get rid of any possible metaphysics and treat an expression and the judgment it represents as identical. For this purpose he needs a new symbolism.[23] The construction of a system of logic and mathematics is the task of constructing a system of pure rules concerning the combination of symbols and letters.[24] This leaves no room for a metaphysics—involving only the ability of the reader to distinguish between symbols and letters. Logic and mathematics become from this point of view a system of signs upon which one operates in definite prescribed ways. For this reason Chwistek calls his system "Nominalistic" or "Semantic." From the semantic point of view, which treats of the expressions, the expressions $\psi\{x\} \vee \psi\{x\}$ and $\psi\{x\}$

---

[20] Cf. Const. Types, pp. 26–27.
[21] Cf. Black, *op. cit.*, pp. 135–37.
[22] Cf. N. G., p. 387.
[23] G. L. M., p. 708.
[24] *Ibid.*, p. 704, and N. G., pp. 367–68.

are clearly different although they are equivalent for all x. Chwistek considers mathematics itself to be an "object language," and his notation attempts to treat it as such.[25]

Chwistek aims to derive a syntactical structure which one can use to build up the *patterns* of the propositions of logic and mathematics. The system is to be purely formal in this sense. No meaning is to be attached to the symbols—they merely represent themselves, and are used to express the way in which these propositions of logic and mathematics are put together; but they must be capable of interpretation. In brief, it is possible to say that what Chwistek is attempting is the construction of the syntactical structure of logic and mathematics. Of course, Chwistek, at the bottom, seems to identify logic and mathematics with their syntactical structure, i.e., he accepts the definition of mathematics in terms of its structure. The system which results is to be viewed entirely as a system of rules for the manipulation of symbols and letters. We have, as a result, a fusion of the symbol, the name of the symbol, and the "meaning" of the symbol—or rather the "meaning" of the symbol is said to be the symbol itself.

It is perhaps appropriate to point out here that although syntax is very important and its analysis certainly a legitimate task, to confuse the meaning of a term with the structure of the term is to confuse a difficult problem. It is as if one were to identify the meaning of "love" with the word itself. The derivation of the structure of mathematical propositions from semantical rules is not the same as the reduction of the meaning of a mathematical proposition to its syntax.

Since Chwistek makes no distinction between an expression and its name, either can be substituted, under given conditions, for the symbols in the formulas. From Chwistek's point of view this is not in itself a confusion, for if what he is giving is a sort of universal meta-language, then it should be equally applicable to symbols and to the names of symbols. The rules of substitution must, however, be so stated that sentences do not confuse the two. The system must provide rules for substitution and for interpretation, both of which Chwistek attempts to formulate in F. M., which is a modification of the approach in G. L. M. but not a change of point of view. We follow F. M. unless otherwise noted.

Chwistek first introduces four propositions of a preliminary language (i.e., meta-language) which will be used in talking about the language in terms of which he will develop his system. This preliminary language is to serve as a sort of machine for turning out symbolic expressions which will make possible the construction of the theorems

[25] Cf. Z. Jordan, "The Development of Mathematical Logic and of Logical Positivism in Poland between the two Wars," Polish Science and Learning, London and New York, 1945, pp. 26–27.

of what Chwistek will call the "auxiliary system." The auxiliary system contains no variables and is used to build up systems containing real and apparent variables. This leads to still higher systems. These propositions of the preliminary language are:[26]

x is an expression.
x is an abbreviation of y.
x is a theorem.
If P then Q.

Two primitive signs, the asterisk (*) and the letter c, are introduced. The * is used as a parenthesis (including exactly two terms) and c is a fundamental element without any meaning in itself. It is apparently used as a symbol to represent whatever at the moment may need representation without specification, whether it be an expression in syntax or in semantics. The * appears to represent an operator at times. If "enclosing in parentheses" is an operation, then the * is always to represent an operation.

Expression is defined recursively as follows:

A_0. c is an expression.

If E is an expression, if F is an expression, then $*EF$ is an expression. This enables us to build up any number of expressions, e.g., A_0 allows us to say $*cc$ is an expression, and since $*cc$ is an expression and c is an expression then $*c*cc$ is an expression.

Abbreviated expressions are defined:

A_1. If E is an abbreviation of F, if G is an abbreviation of H, if I is an expression, then

$*EG$ is an abbreviation of $*FH$
$*IG$ is an abbreviation of $*IH$
$*EI$ is an abbreviation of $*FI$

Certain expressions are introduced as the elements of *significant expressions*. These *significant expressions* are abbreviations of a selected set of expressions.

A_2. If L is an expression, then

$.0(L)$ is an abbreviation of $*LL$
$.1(L)$ is an abbreviation of $*.0(L).0(L)$
$\mathrm{I}(L)$ is an abbreviation of $*.0(L).1(L)$
$\mathrm{II}(L)$ is an abbreviation of $*.1(L).0(L)$

and in particular

0 is an abbreviation of $.0(c)$
1 is an abbreviation of $.1(c)$
I is an abbreviation of $1(c)$
II is an abbreviation of $\mathrm{II}(c)$

[26] F. M., p. 1.

thus 0 is an abbreviation of .0(*c*) which is an abbreviation of **cc*
1 is an abbreviation of .1(*c*) which is an abbreviation of *.0(*c*) .0(*c*) which is the expression ***cc***cc*.

**cc* may be viewed as the pattern of $p|p$ written as $|pp$ and read "*p* is incompatible with *p*"—the Sheffer stroke formula from which all of the logic of Principia can be derived by appropriate definitions.[27] Chwistek used the Sheffer stroke in G.L.M. (p. 528) to show that logic can be derived from his system.

It is to be recalled that these are schematic representations of the structures of statements, where *c* represents an expression and * represents an operation.

Some combinations of these significant expressions may be selected and called *patterns*.

The following directions are given by Chwistek:

1. *a* is an auxiliary letter. If *E* is an auxiliary letter, *E′* is an auxiliary letter.

2. *c* is a pattern. If *E* is an auxiliary letter, *E* is a pattern.
If *E* is a pattern, if *F* is a pattern, then **EF* is a pattern.

3. If *E* is a pattern, if **FG* is a pattern, and if **FG* is contained in *E*, then **FG* is an *element* of *E*.

4. If *E* is a pattern, *I* is an auxiliary letter contained in *E*, *F* is an expression and *G* is the result of substitution of *F* for *I* in *E*, then *G* is a *value* of *E*.
If *G* is a value of *F* and *F* is a value of *E*, then *G* is a value of *E*.
If *F* is an expression and *F* is a value of *E*, then *F* is a constant value of *E*.

(The definition of mutual independence of patterns is omitted).

Let us illustrate these concepts in a loose fashion.

Let us have given the expressions *p*, *q*, *r*, (), →

Then *pq* is an expression as are *pr*, *qr*, *pqr*, $p{\rightarrow}qr$, $\rightarrow pqr$, (), *p*, *r*, etc.

Then *pp* is a significant expression as are *qq*, *rr*, *ppqq*, $p{\rightarrow}qr$, $((p{\rightarrow}q).p){\rightarrow}q$, etc.

Now, these may all also be called patterns, and in particular, we can have the pattern $((p{\rightarrow}q)p){\rightarrow}.q$.

---

[27] Nicod, "Reduction in Number of Primitive Propositions of Logic," *Proc. Cambridge Phil. Society*, *xix*, 1917, pp. 32–41.

The remaining explanations can now be supplied by the reader.[28]

We are now ready to construct what Chwistek calls the *auxiliary system*—which is fundamentally the pattern of the syntax of logic. The fundamental ideas of this system consist of two patterns and a relation between patterns:

1) the pattern of substitution,

2) the pattern of elementary systems,

3) the Sheffer stroke.

First, two abbreviations are given

A_3. $(EFGH)[L]$ is an abbreviation of $****\mathrm{I}(L).0(E).0(F).0(G)$ $0(H)$

A_4. $|EF$ is an abbreviation of $*\mathrm{I}(E)\mathrm{II}(F)$

A_3 is the pattern of substitution and is interpreted by I_3 just below.

Auxiliary propositions can now be built by using the following system of interpretations:

I_1. $*cc$ is an expression of type c.
If E is an expression of type c, and if F is an expression of type c, then $*EF$ is an expression of type c.

I_2. If $****OEFGH$ is an expression of type c then $(EFGH)[c]$ is an auxiliary proposition.
If E is an auxiliary proposition and if F is an auxiliary proposition, then $|EF$ is an auxiliary proposition.

I_3. If $(EFGH)[c]$ is an auxiliary proposition, then "$(EFGH)[c]$ is true" means "H is the result of substituting G for F in E"; "$(EFGH)[c]$ is false" means "H is not the result of substituting G for F in E."

I_4. If $|EF$ is an auxiliary proposition, then "$|EF$ is true" means "either E is false or F is false"; "$|EF$ is false" means "E is true and F is true."

(I_3 is a modification of the symbolism in G.L.M. where Chwistek tried to equate by definition what he called the fundamental schema with Sheffer's stroke. Cf. G.L.M. 1, p. 708 and G.L.M. 2, pp. 527–28. Also compare N. G., p. 374.)

Some illustrations of I_3 are:

1. Let E, G, H be c, then we have $(cFcc)[c]$ which reads "c is the

[28] The point of view of Chwistek has decided affinities with that of W. V. Quine, *Mathematical Logic*, W. W. Norton, 1940, although Quine's stratified languages are not quite the same.

result of substituting *c* for *F* in *c*." Since *c* is a constant expression, the above then represents the pattern of a proposition.

2. Let *E* and *G* be *c*, then "(*cFcH*)[*c*]" reads "*H* is the result of substituting *c* for *F* in *c*." This is the fundamental pattern of "follows from."

Since Sheffer has shown, in the work cited above, how to derive all logic from the stroke function, Chwistek is now in position to do likewise using the following abbreviations and their interpretations (A_5 and I_5).

Abbreviations

$\sim E$ is an abbreviation of $|EE$
$\supset EF$ is an abbreviation of $|E{\sim}F$
$\vee EF$ is an abbreviation of $|{\sim}E{\sim}F$
$\wedge EF$ is an abbreviation of $\sim|EF$
$\equiv EF$ is an abbreviation of $\wedge\supset EF\supset FE$

Interpretation

"$\sim E$ is true" means "*E* is false"
"$\supset EF$ is true" means "if *E* is true *F* is true"
"$\vee EF$ is true" means "either *E* is true or *F* is true"
"$\wedge EF$ is true" means "both *E* and *F* are true"

Other ideas of the semantical calculus can now be built up on these foundations. Using this elementary system, Chwistek now attempts to construct a semantical calculus in which the integers can be constructed. This is done by means of a series of abbreviations as follows:

A_9 $.1(L)$ is an abbreviation of $.0(.0(L))$
$.2(L)$ is an abbreviation of $.0(.1(L))$
. . . .
$.10(L)$ is an abbreviation of $.0(.9(L))$, etc.

From this point on, Chwistek defines various semantical and syntactical principles such as the principle of logical deduction and the principle of substitution.

This is sufficient to indicate the different point of view of Chwistek. Logic and mathematics are what we get when we make certain types of substitutions in the fundamental schema. The types of substitutions are all enumerated so that no paradoxical element can enter. Such a point of view carries the formal aspect of mathematics to its extreme.

Chwistek's work deserves to be studied much more carefully than we have time to do here. Its relation to the general view of the nature of mathematics needs careful analysis. It is an attempt to use a symbolism which makes possible the treatment of both the symbol and the judgment the symbol represents.[29]

Chwistek's work is really a connecting link between the logistic and formalistic points of view. He accepts the logistic thesis yet tries to retain the formalist emphasis on symbolic manipulation by reducing the problem to one of syntax. It is unfortunate that Chwistek did not survive the war to develop his ideas further. Perhaps the above discussion will inspire some student to elaborate the approach outlined in these pages.

---

[29] Two papers of Chwistek indicate some of the more metaphysical implications of rational semantics: "Pluralité des réalities," *Atti del V. Congress. intern. di filosofia*, Naples, 1924; "La Semantique rationelle et ses applications, "*Travaux du IX*e *Congrès International de Phil. VI, Logique et Mathématiques*, Paris, 1937, pp. 77 ff.

CHAPTER NINE

THE FOUNDATIONS OF MATHEMATICS

II. HILBERT'S FORMALISM

The first edition of Hilbert and Ackermann's book that we followed in our discussion of the logistic foundations, ends with the remark "A general structure of logic, free of the difficulties of the axiom of reducibility, is found in D. Hilbert's investigations of the foundations of mathematics, a unified presentation of which will appear later."[1] The extremely close relation between formalism and logistics is implied in these words. However, there is a fundamental difference in point of view between the two approaches. Russell seems to have two points of view—a symbolic technique and a metaphysics of logic (especially a metaphysics of classes). His symbolic technique implies, however, that mathematics, i.e., logic, deals with symbols which can be viewed as pure structures independent of any reference.

This tendency is illustrated further in Chwistek's semantic approach and is adopted explicitly by Hilbert. Mathematics, for Hilbert, is like a game played with *arbitrary* symbols in accordance with arbitrary rules, subject only to the formal condition of consistency. Mathematics is like a game of chess in which you can set up rules of moving the players at will—provided always your rules do not lead to p and p',—i.e., to purely formal contradictions, and enable you to decide of a given combination whether or not it is permissible. Such a position is clearly a nominalistic one.[2] The difficulties of such a position, especially in explaining the application of mathematics, will be discussed later. We must, however, keep in mind that although there is a vital difference between a *theory about mathematics* and a *mathematical* theory, the theory about mathematics must not be such as to make impossible an explanation of the possibility of a mathematical theory. What Hilbert calls "metamathematics" is a theory about mathematics. Thus Hilbert (who is the leader of the formalist school and whose work will for them basis of our discussion) considers mathematical objects as symbols

[1] Hilbert and Ackermann, *Grundzüge der Theoretischen Logik*, p. 115.

[2] Jörgenson, *Treatise of Formal Logic*, Vol. 3, p. 25. But cf. the excellent summary of Hilbert's development given by H. Weyl, "David Hilbert and His Mathematical Work," *Bulletin of American Mathematical Society*, Vol. 50, No. 9, Sept., 1944, pp. 612 ff., especially pp. 635–45.

implicitly defined by systems of axioms. Since symbols are not logical constants, mathematics can never be founded by logic alone.[3] The emphasis of the formalists is laid upon axiom systems and the investigation of their properties.

Logic for Hilbert is a tool which one uses to lay bare the structure of mathematical axioms. It offers us a symbolism, which is a formal language, by which we can formalize not only the structure of mathematics but also itself as well.[4] When we apply the formal language to the analysis of mathematics we are led to the construction of a set of axioms, i.e., to set up a few propositions from which all others of the system can be deduced. But when we do this we find that problems arise concerning the *structure* of mathematics which form a sphere of problems unique in themselves. These are not what are usually called mathematical problems, but rather are problems concerning proofs in mathematics. A new theory about mathematics results. This new theory of proof Hilbert calls *metamathematics*. Metamathematics deals, as Hilbert says, with the proofs of mathematics in the same way as "the physicist investigates his apparatus." Hilbert desires to construct a perfectly rigorous mathematics which will not lead to contradictions or paradoxes. He feels that this can best be done by the means which were so successful in geometry—the axiomatic method. But the axioms must be shown to be consistent, i.e., we must have some proof that we will not be able to deduce *p* and not-*p* from these axioms. Hence Hilbert's interest in the consistency proof.[5] The investigations into the properties of axiom sets has led to outstanding results in logic—different types of consistency and other problems have been uncovered and solved in part at least. As a matter of fact, metamathematics is really logic. Its problems—independence, consistency, categoricity, etc., are problems of logic and are investigated in logic texts.

Hilbert and his followers consider mathematics to be a purely formal calculus—i.e., it is almost a mechanical manipulation of symbols which have no reference to any particular "actual" entities. Mathematics is therefore a pure calculus,[6] and can be replaced by a method, entirely mechanical, for deducing formulae.[7] This does not mean, as is frequently supposed, however, that mathematics for Hilbert is a game with *meaningless* symbols. In an article in *Mathematische*

[3] Hilbert, "Die Grundlagen der Mathematik," *Abh. Math. Semin. Hamburg, Univ.*, Vol. 5, 1927.

[4] Hilbert-Bernays, *op. cit.*, p. 45, also Carnap, *Logical Syntax of Language*, N. Y., 1937, pp. 325 ff.

[5] Cf. H. Weyl, "Consistency in Mathematics," *Rice Institute Pamphlets*, 16, 1929, p. 250.

[6] W. Dubislav, *Die Philosophie der Mathematik*, pp. 47–48.

[7] A. Heyting, *Mathematische Grundlagenforschung*, p. 37. This is an excellent outline of Hilbert's work and its development.

Annalen ("Die Grundlegung der elementaren Zahlenlehre," 1931), he appears to consider mathematics as expressing the *a priori* conditions for any science whatsoever. The symbols may be *arbitrary* but not meaningless. It does mean, however, that the formalist may ignore the application of his system. Only in this sense are the symbols meaningless.

And to be *like* a game is not to be a game. Hilbert's use of the term *thought-thing* (*Monist*, Vol. 15, 1905, p. 341) indicates more than meaningless symbol. Pure axiomatics presupposes a sphere of objects which it presents in an idealized form, says Hilbert in his latest work, *Grundlagen der Mathematik*, Volume 1, pages 2–3. In the early pages of this work a much more metaphysical position is taken than in any other of the works of Hilbert. The application of this method to all knowledge had been emphasized by Hilbert in 1931 (p. 494 cited below) also and he said then " . . . My theory of proof actually is nothing more than the description of the innermost process of our understanding and it is a protocol of the rules according to which our thought actually proceeds" (p. 493). As a consequence, all criticisms of Hilbert based on the idea that he treats of meaningless symbols must be abandoned.[8] But pure mathematics treats of the combinations of these symbols.

Hilbert's approach to the investigations of the number concept is determined by three facts: (1) the dependence of all branches of mathematics upon arithmetic—i.e., upon the idea of number (2), the avoidance of the contradictions and paradoxes which result from the logistic approach; (3) the successful formulation of geometry as an axiomatic structure.[9] This last was accomplished by Hilbert himself in his famous "Foundations of Geometry." We do not consider this here. The axiomatic method, which he advances in a series of articles and whose theory he develops in its latest form in Hilbert-Bernays, *Grundlagen der Mathematik* (Berlin, 1936), Hilbert early felt to be applicable to these investigations which had so disastrously run up against a stone wall.

Chwistek reduces logic and mathematics to deductions by substitution in the schema *EFGHJ. Hilbert feels that already in logic certain arithmetic concepts are used and hence he speaks of the fact that certain notions, as, for example, unity, are common to experience and pure thought when these are viewed as sources of our knowledge of mathematics. Hilbert also calls the attempt to define the numbers purely logically, an erroneous direction. The criticism that any definition of "one" involves the concept of "one" is at times still advanced and is due to the fact that in setting up a correlation between two classes

[8] Cf. A. Müller, "Über Zahlen als Zeichen," *Math. Ann*; Vol. 90, 1923, pp. 153–58 and Birkmeister, *Uber den Bildungswert der Math.*, Leipzig, 1923, pp. 28 ff.

[9] Cf. Hilbert, "The Foundations of Logic and Arithmetic," *Monist*, Vol. 15, 1905, pp. 339 ff.

we need to select *one* member from each class. Thus it is felt that the concept of the number one is "smuggled" in and then derived. This is, however, to confuse the idea of "unity" and the ability to recognize a *unit*, with the concept *the number one*. But if we accept this reply to the vicious circle criticism, we must accept the conclusion that the number one is based on the idea of unity and involves a psychological factor—namely recognition of unity. It is recalled that Chwistek also, in introducing his ideas, said that all that was presupposed was the ability to distinguish different letters and symbols. The philosophical discussions, which we mentioned earlier, concerning quantity were really psychological investigations concerning the concept "unity." It appears rather startling to suggest that perhaps our arithmetic depends upon our psychology. The pure formalist approach leaves such a possibility quite open, even though it refuses to concern itself with the problem.

To return to Hilbert. As a consequence, Hilbert feels that a simultaneous development of logic and arithmetic is needed.[10] This, again, can be done by the axiomatic method. The notion of number is to be developed on the basis of the idea of a *thought-thing* and is not to be accepted in any dogmatic fashion, as Kronecker does.[11] Thus the natural numbers cannot be viewed as given, but are to be constructed by means of combination of these *thought-things*. Hilbert takes a definite stand against the intuitionists (whom we discuss later) although later he comes closer to their viewpoint.[12]

In the article translated by Halsted in the *Monist* (1905), Hilbert outlines his general ideas. We present them before we go on to the more detailed analysis in later articles and especially in the *Grundlagen*, Volume 1.

We consider the thought thing 1 (read "one"—later also written /) and combinations of the thing with itself, e.g.,

11, 111, 1111, 11111, etc.,

as well as combinations of these in any way as, e.g.,

(1)(11), (1)(11)(11), [(11)(11)](11), [(111)(1)(11)](111), etc.

We shall call 1 a *simple* thing.

The ability to *recognize* 11 does not involve the use of the *number 2*. This must be kept in mind.

We now adjoin a second *simple* = (read "equal") and combine it with the combinations of 1.

1=, 11=, ...(1)(=1)(===), [(11)(1)(=)](==), 1=1, (11) =(1)(1).

[10] *Ibid.*, p. 340.
[11] *Ibid.*, pp. 338 and 341.
[12] *Ibid.*, p. 352.

If we have a combination a of things 1 and = and another b, then a differs from b unless the combinations are identical.

We now conceive 1, =, and their combinations to be divided by some means into two classes which we shall call the class of *existents* (any combination of 1, = in this will class be denoted by a) and the class of *nonexistents* (any combination in this denoted by $\bar{a}$). Every combination in the class of existents is different from every one in the class of nonexistents, and every possible combination belongs to one of the classes. $\bar{a}$ also denotes the statement that a belongs to the class of nonexistents, i.e., a denotes a combination of 1, =, and the statement that it belongs to class of *existents.* If the combination does actually belong to the class of existents, then a is said to be a *true* statement. Likewise $\bar{a}$ is true if a belongs to the class of nonexistents, since $\bar{a}$ means "a belongs to class of nonexistents." Clearly then, since every combination in the class of existents differs from every one in that of nonexistents, a and $\bar{a}$ cannot both be true. The statements a and $\bar{a}$ then, constitute a contradiction since $\bar{a}$ means that a belongs to class of nonexistents.

If we have two statements A and B such that "from A follows B" or "if A is true B is also," then we denote this by $A|B$ and call A the *hypothesis* and B the *conclusion.* A and B may be composed of statements A_1, A_2, B_1, B_2, B_3 so that, e.g.,

$$A_1 \& A_2 | B_1 \text{ o } B_2 \text{ o } B_3,$$

i.e., from A_1 and A_2 follows B_1 or B_2 or B_3 (& means "and," o means "or").

Let $A_1, A_2, A_3, \ldots$ stand for those statements which result from a statement $A(x)$ when we replace x by combinations of 1 and =.

$A_1 \text{ o } A_2 \text{ o } A_3 \text{ o} \ldots$ will be denoted by $A(x^{(o)})$ (read "at least for one x").

$A_1 \,\&\, A_2 \,\&\, A_3 \,\&\, \ldots$ will be denoted by $A(x^{(\&)})$ (read "for every single x").

We now write

1. $x = x$

(read, "every x is identical with that x").

2. $[x = y \,\&\, w(x)] | w(y)$

(read "if x is identical with y and $w(x)$, then $w(y)$").
These form the definition of = and, are called *axioms.* Inferences from these are obtained by the insertion for x of combinations of 1 and =.

To demonstrate the consistency of 1 and 2 we proceed as follows. Consider a set of inferences such that the conclusion of the last has as ts hypothesis the conclusion of the prior inference, e.g., $A|B$ and $B|C$. Then if we take $A|C$ we call this an *inference* also. Select those inferences a which are not of form $A|C$ but are simple affirmations. All things a which result are to be put in the class of existents while all others differing from these in the class of nonexistent.

We see from the form of 1 and 2 that only affirmations of the form $a = a$, i.e., belonging to the class of existents, can occur. Hence, the class $\bar{a}$ is empty. The axioms 1 and 2 are then said to be free from contradiction.

We now add new thought-things.

u (infinite aggregate), f(following), f^1 (accompanying operation), and the axioms:

3. $f(ux) = u(f^1x)$
4. $f(ux) = f(uy) \,|\, ux = uy$
5. $\overline{f(ux) = ui}$

where ux denotes an element of the infinite aggregate u, and $f(ux)$ is a new thought-thing defined by the element x of u.

Above, 3 reads then: "$f(ux)$—the new thought-thing following x in u—is the same as the element f^1x of u."

Also, 4 reads: "If the thought thing $f(ux)$ following x in u is equal to the thought thing $f(uy)$ following y in u, then the elements x and y are identical."

Likewise, 5 reads: "The statement '*the element* $f(ux)$ *following* x *in* u *is the element* 1 *of* u' belongs to the class of nonexistents." This means 1 is the first element of u.

We must make sure, again, that we do not get from these new axioms statements of form a and $\bar{a}$. Now from the form of the axioms we see that only 5 can yield a statement $\bar{a}$. Therefore, if a contradiction is to arise we must be able to infer from the axioms

6. $f(ux^{(0)}) = u1$

i.e., the thing $f(ux)$ following some x in u is the element 1 of u. We must show this cannot be derived.

Let us call every statement of form $a = b$, where a and b are both combinations of same number of simple things—e.g., $(f1)(f1) = (1111)$ and $[f(ff^1u)] = (1uu1)$—*homogeneous equations*. Obviously from 1–3 only homogeneous equations can result as conclusions. In 4, if the hypothesis is homogeneous, the conclusion will also be. Equation 6, if it results at all, must result from 1–4. But 6 is not homogeneous and cannot result if in 4 we assume the hypothesis homogeneous. Therefore the set is consistent. This proof is obviously incomplete and demonstrates the consistency of the set only under definite limitation.

In this article Hilbert also explicitly states the position of the formalists regarding existence. He says concerning the aggregate (infinite) *u* that the affirmation of its existence is justified by the demonstration of consistency. As a matter of fact, a statement is true or is an admissible postulate if it can be shown that it leads to no contradiction. The importance of consistency is emphasized strongly as a requisite in any method of establishing the foundations of mathematics.[13]

The article just discussed contains the germ of all formalist ideas: (1) that an axiomatic system is the only sure way of establishing logic and mathematics; (2) that mathematics cannot be derived from logic but the two are to be simultaneously developed; (3) that the introduction and use of new concepts is always justified by consistency proofs; (4) the investigation of the consistency problem is indicated.[14]

In 1922–23 two papers by Hilbert appeared which evidenced the progress he was making in developing this theory. The first was a lecture held at Hamburg in which Hilbert distinguished between types of symbols and gave a consistency proof of a set of given axioms,[15] which depended upon the proof of a lemma which he was not able to give. The second was a sketch of his ideas which he now says have as their fundamental purpose the removal of all uncertainty regarding mathematical deduction.[16] In this paper also, Hilbert comes to a full recognition of the distinction between a mathematical theory and a theory about mathematics which he calls *metamathematics.* Hilbert feels he has developed this theory about mathematical proof to such a degree that one can obtain by it a foundation for analysis and the theory of aggregates against which no objection can be raised. In the article "Logische Grundlagen," Hilbert gives an outline of his ideas to that date.

All mathematics is to be formalized so that mathematics proper (or mathematics in the narrower sense) is put into the form of formulae. The formulae differ from those of ordinary mathematics by the occurence of logical symbols, especially that for "follows" (→) and that for "not" (—). Certain formulae which serve as the foundation for the

---

[13] Hilbert had already proved the consistency of the axioms of Geometry (Euclidean). Cf. *Foundations of Geometry,* trans. Townsend, 2nd ed., Chicago, 1921, pp. 27 ff.

[14] Cf. also Hilbert, "Axiomatisches Denken," *Math. Ann.,* Vol. 78, 1918, pp. 405 ff.

[15] Hilbert, "Neubegründung der Math.," *Abhand. aus dem math. Seminar der Hamburgischen Universität,* 1922, Vol. 1, p. 157. Cf. L. O. Kattsoff, "Postulational Methods I," *Phil. of Science,* 1935, Vol. 2, pp. 147–49.

[16] Hilbert, "Die Logischen Grundlagen der Mathematik," *Math. Ann.,* Vol. 88, 1923, pp. 151 ff.

formal structure are called axioms. A *proof* is a form which must be given intuitively. It has the schema;

$$\frac{\begin{array}{c} S \\ S \rightarrow T \end{array}}{T}$$

where the premises S and $S \rightarrow T$ are either axioms, or are obtained from axioms by direct substitution, or are the T of some other such schema. This schema has properties analogous to the principle of the syllogism. If a formula is an axiom or derived from an axiom in the ways mentioned, it is said to be *demonstrable* (beweisbar).

The theory which concerns itself not with deriving proofs but with the properties of proofs as such, is called *metamathematics.*

The axiom system for the theory of proof, given in "Die logischen Grundlagen der Mathematik," is:

I. Axioms of deduction:

1. $A \rightarrow (B \rightarrow A)$
2. $\{A \rightarrow (A \rightarrow B)\} \rightarrow (A \rightarrow B)$
3. $\{A \rightarrow (B \rightarrow C)\} \rightarrow \{B \rightarrow (A \rightarrow C)\}$
4. $(B \rightarrow C) \rightarrow \{(A \rightarrow B) \rightarrow (A \rightarrow C)\}$

II. Axioms of negation:

5. $A \rightarrow (\bar{A} \rightarrow B)$ (Hilbert calls this "law of contradiction.")
6. $(A \rightarrow B) \rightarrow \{(\bar{A} \rightarrow B) \rightarrow B\}$ (Principle of excluded middle. The universal applicability of this axiom is denied by the intuitionists)

III. Axioms of equality:

7. $a = a$
8. $a = b \rightarrow \{A(a) \rightarrow A(b)\}$

IV. Axioms of number:

9. $a + 1 \neq 0$ (this is the same as $\overline{a + 1 = 0}$.)
10. $\delta(a + 1) = a$

V. Transfinite axiom:

11. $A(tA) \rightarrow A(a)$

This can be read: (tA) is an entity such that if it has the property A, then all a have that property. (This axiom should be noted since it forms another point contested by the intuitionists. Hilbert feels justified in using this if he can show it consistent with other axioms.) An example of this axiom is: if tA is a triangle and we can predicate of this triangle that its angle sum is 360°, then we can predicate of *all* triangles that their angle sum is 360°.

VI. Definitional-axioms for "all" and "there exists" symbols:

$A(tA) \rightarrow (a)A(a)$
$(a)A(a) \rightarrow A(tA)$
$A(t\bar{A}) \rightarrow (Ea)A(a)$
$(Ea)A(a) \rightarrow A(t\bar{A})$

It is now necessary to show the consistency of axioms 1 to 11. The fundamental idea behind the proof is to show that no proof can be constructed which has as its final formula (T) a formula of the type $0 \risingdotseq 0$. This is done by showing that the assumption that $0 \risingdotseq 0$ does occur, involves a contradiction. This really involves the law of contradiction, hence if this law is taken as an axiom in the system, the proof is circular. Since consistency is demonstrated by proving that p and not-p cannot both be derived in the system, and this is really the assertion of the law that p and not-p cannot both be true, any system that contains the law could not be proved consistent in this sense without a circular argument. This implies that for the consistency proof of arithmetic we need the laws of logic—but for that of logic we need another system or method. The method of demonstrating the impossibility of $0 \risingdotseq 0$ occurring, proceeds by the evaluation of the proof in such a way that all variables are eliminated, and in every formula only the symbols 0, $0 + 1$, $0 + 1 + 1$, etc., and logical symbols occur. Every formula is thus reduced to a "normal" form. We can then determine for every formula whether or not it is *true*. If it can be shown in general that the supposed (or purported) proof for $0 \risingdotseq 0$ contains all true formulae, then the proof cannot end in $0 \risingdotseq 0$. This process must be completed in a *finite* number of steps.

Of especial importance is the consistency of the transfinite axiom, since this justifies the use of transfinite methods of deduction in mathematics. Only the most simple case of this will be presented in more detail. This occurs if in axiom 11 we take for objects a the symbols for positive whole numbers including 0, and as predicate $A(a)$ the equations $f(a) = 0$, where f is an ordinary whole-numbered function. The logical function t correlates a number to every f. From t there arises an ordinary whole-numbered function of a function such that if f is a definite function, t is a definite number, which we denote by $t(f)$.

$$t(f) = t_a[f(a) = 0]$$

Axiom 11 changes for this case to

12. $f[t(f)] = 0 \rightarrow f(a) = 0$

(The property of $t(f)$ can be made clearer by taking $t(f)$ as 0 whenever $f(a) = 0$ for every a, otherwise as the first number a such that $f(a) \neq 0$ whenever $f(a) \neq 0$.)

To show now that axioms 1–10 and axiom 12 are consistent we must extend the consistency proof of 1–10 so as to cover 12.

We now define the following operations:

We assume temporarily that in $t(f)$ only such ϕ can be substituted for f as allow us to change the proof into one containing logical symbols, number symbols, and $t(\phi)$ only.

1. We, in a sense experimentally, replace $t(\phi)$ by 0 everywhere. Our proof then becomes a sequence of "numerical" formulae. All these formulae are "true" except possibly those derived from 12. But from 12 if we replace f by ϕ, make the proper substitution for a, and then replace $t(\phi)$ by 0, we get

$$\phi[t(\phi)] = 0 \rightarrow \phi(z) = 0 \text{ and therefore}$$

$$\phi(0) = 0 \rightarrow \phi(z) = 0$$

Since z is a number symbol and ϕ a function defined by a recursive process[17] $\phi(z)$ is itself always a number symbol. The question then is: Is $\phi(z)$ the number symbol 0 or some other number symbol? If it is 0, then we have a consistency proof since $0 \neq 0$ could not occur.

2. If it is not 0 then $\phi(z) = 0$ is false and we try another process. We replace $t(\phi)$ by the number-symbol z wherever $t(\phi)$ occurs. We then derive from 12, formulae of the form

(a) $\phi(z) = 0 \rightarrow \phi(s) = 0$

Recalling that $p \rightarrow q$ means $\bar{p} \vee q$ and setting up the matrix (cf. Chapter 8), we see that a false proposition implies any proposition, then since $\phi(z) = 0$ is false, the entire implication (*a*) is true and cannot give rise to proofs ending in $0 \neq 0$. (That is, $[\phi(z) = 0 \rightarrow \phi(s) = 0]$ cannot imply $0 \neq 0$ since then $[\phi(z) = 0 \rightarrow \phi(s) = 0] \rightarrow 0 \neq 0$ would be a case of a true proposition implying a false one—which is false.).

This, says Hilbert,[18] is the complete proof of the consistency of the transfinite function $t(f)$—hence we can use it without fear of any inconsistency. Our thought is finite and insofar as we think, a finite process takes place. But this does not invalidate all procedures where an infinite number of objects are involved, since the consistency proof makes it possible for us to act as if we had selected the necessary object without risk of error. The application of the law of excluded middle is always valid. There are of course, difficulties in the proof. It appears to violate the theory of types.

---

[17] The schema of a recursive definition is $f(0) = a, f(n^1) = b[n, f(n)]$. Cf. Hilbert-Bernays, *Grundlagen der Mathematik*, Vol. 1, pp. 286–87.

[18] Hilbert, *Logische Grundlagen der Math.*, pp. 160–61.

(This is the position to which we shall later oppose the *finitist* point of view which will insist that the law of excluded middle—every proposition is either true or false—cannot be applied where no process is described for deciding which alternative is present.)

The importance of the law of excluded middle (*tertium non datur*) becomes one of the outstanding facts of the formalists investigations. The intuitionist challenge of this law (which we discuss later) for use with infinite sets of objects made the issue an important one. Briefly, the intuitionists claim that if we have given a function $f(x)$ and (say) the set of points between 0 and 1, then if we ask whether all points satisfy the function or if there exists an x_i such that $f(x_i)$ is false, then unless we can *actually* select the x_i such that $\sim f(x_i)$, we cannot say that it does or does not satisfy $f(x)$. The question naturally arises whether the use of this law is *consistent* with the other axioms of Hilbert's system. This means that Hilbert will insist that the consistency of the law of excluded middle will give him the right to use it everywhere. This will be his reply to the objection that the use of the law of excluded middle must be restricted (1) to finite sets, and (2) to constructible infinite sets. But a difficulty at once arises. We must be careful not to use the law in proving its consistency; in other words, we must not attempt to demonstrate that the axioms including the tertium are consistent by showing that the assumption of an inconsistency leads to a contradiction. It is necessary to have a new process. An attempt was made by W. Ackermann[19] to carry out such a proof but not in a completely satisfactory fashion.[20]

The difficulties in Ackermann's proof were pointed out by J. von Neumann, to whose paper we now turn.[21] This will indicate to us the advances in the formalist position up to that date. However, it also really fails to give a complete proof for the consistency of the number theory. Von Neumann's paper divides into two parts: (a) a description of the formalistic tools necessary for an investigation of mathematics; (b) the axiom set for mathematics and the attempt to prove the consistency of this set.

The concepts of mathematics can be divided into three classes:

1. Positive whole numbers.
2. Functions of different orders. The first order functions have arguments (one or more) which are the numbers 1, 2, 3, ... Then we have functions whose arguments (one or more) are variables or functions of lower order.
3. Propositions.

[19] W. Ackermann, "Begründung des 'tertium non datur' mittels der Hilbertschen Theorie der Widerspruchsfreiheit," *Math. Annalen*, Vol. 93, 1925, pp. 1–36.

[20] Heyting, *Math. Grundlagen*, pp. 44–45.

[21] J. Von Neumann, "Zur Hilbertschen Beweistheorie," *Math. Zeitschrift*, Vol. 26, 1927, pp. 1–46.

Real numbers (defined in terms of cuts) belong to class 2 since they are functions of an argument which is a variable.

The given division corresponds to Ackermann's division of the combinations of symbols into functionals (Hilbert defines a functional as consisting of the following: (1) a number symbol, (2) a basic variable, (3) an individual, (4) a variable function whose empty places are to be filled by number symbols, basic variables or functions, and (5) functions of functions), functions, and formulae (i.e., two functionals between which the symbol = or ≒ is placed). Von Neumann, however, makes no such distinction and will treat all combinations of symbols as *formulae*.

Simple symbols are divided as follows:

1. Variables x_m where m denotes any number.
2. Constants C_m where m denotes any number.
3. Operations $0_m^{(n)}$ where m and n denote numbers.
4. Abstractions A_m where m denotes any number.
5. Parentheses (and) as well as the comma (,).

Von Neumann emphasizes the extremely formalist approach by insisting that these symbols denote nothing, represent nothing and have only as much significance as the figures on a chessboard. They are merely *analagous* to the concepts of logic and mathematics[22] and these concepts will from time to time be substituted for them. Only in this sense can we say that the symbols *mean* the concepts of logic and mathematics. Interpretations of these symbols are arrived at from some other point of view. They may be helpful but in principle they have no intrinsic relation to the symbols.

Class 1 is familiar to all. Class 2 contains the symbols for concepts which are definite and can be used alone without reference to other concepts. Class 3 contains symbols which can only be used in relation to other symbols, e.g., →, negation ∼, ∨, &, +, ., =. The upper index in $0_m^{(n)}$ denotes the number of concepts needed to apply the operation. For example, if $0_m^{(n)}$ is ∼, then (n) is 1; for →, ∨, &, +, etc., it is 2. In other words the index (n) denotes whether the operation is monadic, dyadic, etc. Class 4 contains those symbols whose meaning is closely tied up with that of "variable." Such symbols restrict the substitutions possible for variables and actually "bind" the variable. The term "abstraction" which von Neumann uses refers to what are now called "quantifiers." Such are, for example, the existential quantifier "there exists an x" and the universal quantifier "for all x."

---

[22] Von Neumann, *op. cit.*, pp. 4–5.

The objects of formalistic mathematics are certain combinations of the simple symbols which we call *formulae*. These are defined inductively as follows:

1. (a) Every variable x_m is a formula (m definite).
   (b) Every constant C_m is a formula (m definite).

2. (a) Let $0_m{}^{(n)}$ (m and n definite) be an operation and $a_1\, a_2 \ldots a_n$ be any combinations of simple symbols already known to be formulae, then $0_m{}^{(n)}(a_1\, a_2 \ldots a_n)$ is a formula.
   (b) Let A_m (m definite) be an abstraction and x_n (n definite) a variable. Let a be a combination of simple symbols known to be a formula.
   $A_m{}^{(x_n)}(a)$ is also a formula.

This definition enables us to construct the formulae successively.

Certain transformations are now introduced to make the symbolism more analogous to the usual symbolism.

1. We retain the right to give explanations of the following type: "The constant C_m (m definite) is transformed into Γ" where for Γ any arbitrary symbol may be used: Then *wherever* C_m occurs, Γ must replace it. Abstractions A_m may be transformed in a similar fashion.

Two different C_m, however, must not be transformed into the same Γ. The Γ must not be one of the original simple symbols.

2. We retain the right to give explanations of the following type: "Operation $0_m{}^{(n)}(a_1\, a_2 \ldots a_n)$ passes into $\Gamma(a_1\, a_2 \ldots a_n)$ or $(a_1\, a_2 \ldots a_n)\Gamma$ or $(a_1\Gamma_1\, a_2\Gamma_2\, a_3\Gamma_3 \ldots A_{n-1}\Gamma_{n-1}\, a_n\Gamma_n)$.

3. The use of parentheses:

   a) A formula (a) standing alone may be written a. If a is used to construct new formulae the parentheses must be replaced.
   b) If two parentheses occur in succession one may be omitted. Thus $((a))$ may be written (a).
   c) Where we have $0_m{}^1(a)$ we leave out the parentheses: thus for $0_m{}^1(a_1)$ we write $0_m{}^1 a_1$ also for $\Gamma(a_1)$ we write Γa_1.

We define a *normal formula* as one in which no free variable x_p occurs.

Substitution is also defined. Let a, b be formulae and x_p a variable. If everywhere in a, in which x_p occurs and is free, we replace x_p by b, we clearly get a combination of simple symbols which is also a formula. We denote this new formula by $subst\begin{pmatrix} x_p \\ b \end{pmatrix} a$. Substitution is not an $0_m{}^{(n)}$. This can be extended.

$$\text{Subst}\begin{pmatrix} x_{p_1} x_{p_2} \ldots x_{p_n} \\ b_1 \quad b_2 \ldots b_n \end{pmatrix} a$$

With these formalistic tools we are ready to proceed to define a *proof* and *consistency.*

We transform $0_1^{(1)}(a_1)$ into $\sim a_1$ and $0_1^{(2)}(a_1,a_2)$ into $a_1 \rightarrow a_2$. (These correspond to "negation" and "implication.")

Let R be a definite process by means of which normal formulae can be constructed. We call the process the "*rule of axioms*" and the formulae *axioms.*

If now a rule of axioms R is given, then the *demonstrability* of normal formulae is defined inductively:

(1) If the normal formula a is an axiom, it is demonstrable.

(2) If the two normal formulae a and $a \rightarrow b$ are known to be demonstrable, then the normal formula b is also. This rule will enable us to *construct* demonstrable normal formulae. But if we have given a normal formula it will not in general enable us to decide whether or not the formula is demonstrable. In some cases it might be possible to make the decision but that will be because of special circumstances. This problem of deciding whether or not a given normal formula is demonstrable is the problem of *decidability*—the German is used, "Entscheidbarkeit." So far as von Neumann is concerned, he takes the point of view that in general we cannot decide the question. All we can do is to construct new normal formulae and thus decide for these that they are demonstrable.

An axiom rule R is said to be *consistent* if the two formulae a and $\sim a$ are not both demonstrable for the same R. This merely means that of the four cases:

1. Neither a nor $\sim a$ has been shown to be demonstrable.
2. a has been shown to be demonstrable, but $\sim a$ has not.
3. $\sim a$ has been shown to be demonstrable, but a has not.
4. a and $\sim a$ have both been shown to be demonstrable.

Case 4 can never occur, while 1, 2, and 3 may.

We shall now be concerned with what we shall call the mathematical rule of axioms:

There are given definite schemata, i.e., definite combinations of single symbols, of the symbols a, b, $c \ldots, x_p, x_q, \ldots$, and subst., such that if we replace a, b, $c, \ldots$ by formulae and $x_p x_q \ldots$ by variables and then carry out the substitution, a normal formula results.

We now give an axiom schemata from which if we carry out the substitution just described we derive axioms. These axiom schemata von Neumann insists are purely formal and have no meaning or reference.

I. FINITE GROUPS

A-Group (Logic)

Normal formulae are to be substituted for a, b, c, in these schemata:

1. $a \rightarrow (b \rightarrow a)$
2. $(a \rightarrow (a \rightarrow b)) \rightarrow (a \rightarrow b)$
3. $(a \rightarrow (b \rightarrow c)) \rightarrow (b \rightarrow (a \rightarrow c))$
4. $(a \rightarrow b) \rightarrow ((b \rightarrow c) \rightarrow (a \rightarrow c))$
5.[23] $a \rightarrow (\sim a \rightarrow b)$
6. $(a \rightarrow b) \rightarrow ((\sim a \rightarrow b) \rightarrow b)$

B-Group (Equality)

Transformation: $0_2^{(2)}(a_1 a_2)$ is transformed to $a_1 = a_2$.

For a, b normal formulae, for x_p any variable, for c a formula in which at most x_p occurs as a free variable are, respectively, to be substituted.

1. $a = a$
2. $(a = b) \rightarrow (a \rightarrow b)$
3. $(a = b) \rightarrow (\text{Subst}\binom{x_p}{a}c = \text{subst}\binom{x_p}{b}c)$

C-Group (Non-negative whole numbers)

Transformations: C_1 is transformed to 0. $0_2^{(1)}(a_1)$ and $0_3^{(1)}(a_1)$ to Za_1 or $(a_1) + 1$.

Any normal formulae are to be substituted for a, b.

1. $Z0$
2. $Za \rightarrow Za+1$
3. $\sim((a)+1 = 0)$
4. $((a)+1 = (b)+1) \rightarrow (a = b)$.

(This is essentially the first form of Peano's axioms.)

II. THE T-GROUP

D-Group ("All" and "there exists")

Transformation: The abstraction A_1 is transformed into t; A_2 into Π and A_3 into Σ. The Π is the concept "all" and the Σ "there exists."

[23] This axiom enables von Neumann to give a definition of consistency equivalent to the one he uses. If both p and $\sim p$ are demonstrable, then any formula whatever will follow; if, therefore, it can be shown that not every formula is deducible, then it follows that p and $\sim p$ are not both deducible. Hence an axiom rule may be shown consistent by showing that not every formula is deducible.

For b a normal formula, for x_p any variable, for a a formula in which at most x_p is free, are, respectively, to be substituted.

1. $\Pi^{(x_p)}(a) \rightarrow \mathrm{Subst}\begin{pmatrix} x_p \\ b \end{pmatrix} a$
2. $\mathrm{Subst}\begin{pmatrix} x_p \\ b \end{pmatrix} a \rightarrow \Sigma^{(x_p)}(a)$
3. $\mathrm{Subst}\begin{pmatrix} x_p \\ t^{(x_p)}(a) \end{pmatrix} \rightarrow \Pi^{(x_p)}(a)$
4. $\Sigma^{(x_p)}(a) \rightarrow \mathrm{Subst}\begin{pmatrix} x_p \\ t^{(x_p)}(\sim a) \end{pmatrix} a$

E-Group (Functions)

Transformation: $0_3^{(2)}(a_1,a_2)$ transforms to $\phi(a_1,a_2)$

We can substitute for x_p,x_q two different variables and for a a formula in which x_q at most occurs as a free variable.

1. $\Sigma^{(x_p)}(\Pi^{(x_q)}(\mathcal{Z}x_q \rightarrow (\phi(x_p,x_q) = a)))$

This formula tells us that every expression in the realm of "numbers" can be represented by a function. It may be read as follows: There exists an x_p such that for all x_q, if x_q is a number, then the expression $\phi(x_p x_q)$ represents an operation which establishes the function-value a from the "function" x_p and the "argument" x_q.

III. DEFINITION GROUP

The above five groups are sufficient for the foundation of mathematics but we still need to insert the possibility for the construction of definitions.

Consider the constants, operations and abstractions still unused:

C_m for $m \geqq 2$
$0_m^{(n)}$ for $n = 1$, $m \geqq 4$; or $n = 2$, $m \geqq 4$; or $n \geqq 3$, $m \geqq 1$
A_m for $m \geqq 4$

Obviously there is an infinite series of these. They can be represented in the form of sequences with two indices: the C_m for $m \leqq 2$ by Γ_{uv}; the $0_m^{(n)}$ by $\Omega_{uv}^{(n)}$, the A_m by A_{uv}.
u is called the index and represents the subscript, while v is called the degree and represents the superscript. For example, 0_{uv} represents the series of operators which are obtained when u and v take on specific values from the series 1, 2, 3, ... If $u = 1$, $v = 2$, then 0_{12} represents 0_1^2 which when applied to a_1 and a_2 becomes $0_1^2(a_1,a_2)$ or concretely $a_1 \rightarrow a_2$. That is 0_1^2 is the operator $\rightarrow$.

F-Group (Definitions)

Transformations: Those indicated above into Γ_{uv}, $\Omega_{uv}^{(n)}$ A_{uv}

For a_1, a_2,... normal formulae are to be substituted, for x_r any variable; for a a formula in which at most x_r occurs as a free variable. By $\binom{x_p}{x_q}c$ is meant the formula which results from c when x_p is replaced by x_q in c.

1. $\Gamma_{uv} = a_{uv}$

2. $\Omega_{uv}^{(n)}(a_1\ a_2 \ldots a_n) = \text{Subst}\begin{pmatrix} x_{p_1}\ x_{p_2} \ldots x_{p_n} \\ a_1\ a_2 \ldots a_n \end{pmatrix} b_{uv}^{(n)}$

 where $x_{p_1}\ x_{p_2} \ldots x_{p_n}$ are the free variables in $b_{uv}^{(n)}$

3. $A_{uv}^{(x_r)}(a) = \text{Subst}\begin{pmatrix} x_p \\ (a) \end{pmatrix} \begin{bmatrix} x_b \\ x_r \end{bmatrix} c_{uv}\ (r \neq q)$

This is a description of the process of definition by identity in mathematics. We can, for example, use this to define $3 = ((0 + 1) + 1) + 1$ as follows: $((0 + 1) + 1) + 1$ is clearly equal to some a_{1v}. We then transform the appropriate Γ_{1v} into 3. Likewise $\sim(x_1 \rightarrow \sim x_2)$ is obviously equal to some $b_{1v}^{(2)}$. We then transform $\Omega_{1v}^{(2)}(a_1\ a_2)$ into a_1 & a_2. Also $(x_1 \rightarrow x_2)$ & $(x_2 \rightarrow x_1)$ is equal to some $b_{2v}^{(2)}$. We transform the corresponding $\Omega_{2v}^{(2)}(a_1\ a_2)$ into $(a_1 \leftrightarrows a_2)$.

This gives the complete description of the formalist foundations of mathematics. The next task is obviously to demonstrate the consistency of the set. This von Neumann does not succeed in doing except for the groups *A* to *D*, and *F*. No consistency proof is offered for the entire set although von Neumann believes it attainable.

We shall consider the groups *A*, *B*, and *C* only since these will show us von Neumann's method of consistency proof. More about consistency proofs will be given later, in our discussion of postulational methods in general.

An axiomatic rule is said to be consistent if we have a proof that the two normal formulae a and $\sim a$ are not both demonstrable. Thus the axiomatic rule *R* is consistent if we can divide all normal formulae into two classes *T* and *F* ("true" and "false") having the following properties:

We have a rule which enables us to decide for every given normal formula whether it belongs to *T* or to *F*.

Also: I. $\sim a$ belongs to *T* if and only if a belongs to *F*.

II. $a \rightarrow b$ belongs to *T* when and only when either a belongs to *F* or b to *T*.

III. If a is an axiom it belongs to *T*.

Such a division is called an evaluation (Wertung) and gives a consistency proof—since by II and III every demonstrable formula belongs to T and by I since a and $\sim a$ cannot both belong to T, one of them is not demonstrable. Von Neumann uses a narrower definition, however, for demonstrating the consistency of the A-C group. He calls it a *partial*-evaluation and defines it as follows:

"Given any system S having a finite number of axioms. The rule enables us to divide all normal formulae of S into two classes, T_s and F_s. It also enables us to decide whether a given normal formula belongs to T_s or F_s. And

I′. $\sim a$ belongs to T_s when and only when a belongs to F_s.

II′. $a \rightarrow b$ belongs to T_s when and only when a belongs to F_s or b to T_s.

III′. If a is an axiom of S, it belongs to T_s." (Note that both the valuation and partial-valuation assume the possibility of deciding whether a normal formula belongs to T or F. This means they presume the Entscheidungsproblem to be solved. Where this is not solved this method fails.) The partial evaluation consists in taking a partial group of axioms out of the whole set, for which it is possible to tell whether or not a given formula is demonstrable, and proving its consistency. The procedure then is to take various partial groups of axioms. If all partial groups are consistent, the system is also.

If any two constants, say C_2, C_3, are changed to W_t and W_f, respectively, then the division for Group A-C is defined inductively:

1. (a) Every variable x_m belongs to T.
   (b) Every constant C_m belongs to T except W_f which belongs to F.
2. (a) Let $0_m{}^{(n)}$ be any operation and $a_1\ a_2 \ldots a_n$ any arbitrary formulae, each of which is known to belong either to T or to F. Then
      (1) If $0_m{}^{(n)}(a_1\ a_2 \ldots a_n)$ is identical with $a_1 \rightarrow a_2$, then it belongs to T when and only when a_1 belongs to F or a_2 to T.
      (2) If $0_m{}^{(n)}(a_1\ a_2 \ldots a_n)$ is identical with $\sim a_1$, then it belongs to T if and only if a_1 belongs to F.
      (3) If $0_m{}^{(n)}(a_1\ a_2 \ldots a_n)$ is identical with $(a_1 = a_2)$, then it belongs to T if and only if a_1 and a_2 are identical.
      (4) If $0_m{}^{(n)}(a_1\ a_2 \ldots a_n)$ is identical with $Z a_1$, then it belongs to T if and only if formula a_1 has the form $((\ldots(0 + 1)\ldots) + 1) + 1$.
      (5) If $0_m{}^{(n)}(a_1\ a_2 \ldots a_n)$ is not identical with any of the operations 1 to 4, it belongs to T.
   (b) Let A_m be any abstraction, x_n any variable, and a a formula known to belong to T or to F. Then $A_m{}^{(x_n)}(a)$ always belongs to T. This division demonstrates the consistency of group A-C.

It is advisable to point out that the formalist point of view as described by von Neumann really is applicable to any field of investigation. All it needs is the proper interpretation of the constants, variables, and abstractions and their proper combination to form the axioms of the systems. More of this is in the discussion of the nature of axiom systems.

One thing becomes more and more evident and that is the close interdependence between mathematics and logic. The attempt to establish mathematics on a firm foundation seems to have led to a host of problems for logic to solve. It appears that this might be explicable on the view of mathematics as the science of objects as such. The technique of mathematics is a logical one and it is no wonder that mathematics and logic are so intertwined. By the very definition of consistency, the concept of implication is fundamental. But problems about implication are logical problems. The question at once arises—is logic sufficiently developed to offer us such tools as are necessary to demonstrate consistency? Or perhaps logic can never answer the problem at all. Hilbert's point of view would seem to imply that "whether or not we know a given formula to be demonstrable" is due to our ignorance, but the formula is or is not deducible, and if we sought we could discover the answer. Recent investigations, however, seem to indicate that in any system there will be formulae for which we cannot decide within that system.

To return, however:

In 1930, von Neumann read a paper at Königsberg[24] on the formalist foundations of mathematics in which he stated the problems of Hilbert's theory of proof to be as follows:[25]

1. To enumerate all symbols used in mathematics and logic. These are called "fundamental symbols."
2. To denote uniquely all combinations of these symbols which occur as "meaningful" propositions in classic mathematics. These are called "formulae."
3. A process of construction must be given which enables us to set up all formulae which correspond to the "demonstrable assertion of classic mathematics."
4. It must be demonstrated (in a finite combinatorial fashion) that those formulae which correspond to calculable arithmetics can be demonstrated in accordance with 3, if and only if the actual "calculation" of the mathematical assertions corresponding to the formulae results in the validity of the assertions.

---

[24] Published as "Die formalistische Grundlegung der Mathematik," *Erkenntnis*, Vol. 2, 1931, pp. 116 ff.

[25] *Ibid.*, p. 118.

He then asserts that as a result of the work of the logistic school, 1–3 offer no further difficulties. Only in 4 do we have a real problem. This problem is the finite-combinatoric proof of consistency. And "the present state of affairs" is that the consistency of classic mathematics is still not demonstrated.

In 1934 there finally appeared the first volume of Hilbert and Bernay's book, *Grundlagen der Mathematik*,[26] which gives an authentic form to the formalist position to that date in the light of important investigations. The book is not altogether satisfactory from a pedagogical point of view, but the reader should by all means examine it carefully. The first two chapters are preparatory. Chapters 3 and 4 cover the same ground as the logic. The remainder of the book is concerned with the technical development and detailed proofs.[27]

One defect in the formalist position must be pointed out. A set of axioms can be set up for a given sphere only after we have a great many facts about the sphere. This is true particularly for a set of axioms about a content, e.g., a set of axioms for psychology. This means that the axiomatic approach is a sophisticated position and does not really explain the origin of the concepts involved. Even Hilbert must assume we *know* the thought-object 1. The usual numbers are considered as abbreviations for figures of the form 1, 11, 111, etc. But involved in this act of constructing these successive figures is a whole process of psychological activity. One has the feeling that the number concept is already assumed in identifying 2 with 11, 3 with 111, 4 with 1111, etc. At most (and this is a very important "most") the formalist analysis of a range can demonstrate what is involved in our knowledge. No matter how formalism tries to avoid reference to any content it is inevitably based upon a content. In the case of the foundations of mathematics—i.e., in the case of the numbers—its analysis of them presupposes our ability at least to recognize them. The position of the intuitionists recognizes this and asserts outright that the numbers are given us immediately in intuition.

---

[26] Berlin, 1934. In 1938 a 2nd edition of Hilbert and Ackermann's *Grundzuge der Theoretischen Logik*, appeared which has some important modifications in the logical apparatus.

[27] An excellent summary of Hilbert's work is given by H. Weyl, "David Hilbert and His Mathematical Work," *Bull. Amer. Math. Soc.*, Vol. 50, 1944, pp. 612 ff.

CHAPTER TEN

CHURCH'S ELEMENTALISTIC MATHEMATICS: COMBINATORY FOUNDATIONS

The development of the formalistic investigations has given rise, as was pointed out, to what appears to be almost a branch of mathematics. The problems investigated are borderline problems. But the methods of investigation are mathematical. The best way to gain a notion of these matters is to describe the work of A. Church. Although Church has published a series of works in American mathematical journals[1] we have an excellent introduction to the work of Church in the mimeographed lectures given by him at Princeton in the period October, 1935–January, 1936. It is this work that we shall follow here. We shall give only a bare outline, omitting many theorems and proofs of theorems. Papers of Church which we have drawn upon in this presentation are:

"A Set of Postulates for the Foundation of Logic," *Annals of Mathematics*, Second Series, Vol. 33, No. 2 (1932), pp. 346–66, (cited as F.L. 1).

"A Set of Postulates for the Foundation of Logic (Second Paper)," *Annals of Mathematics*, Second Series, Vol. 34, No. 4 (1933), pp. 839–64, (cited as F.L. 2).

Mathematical Logic (Mimeographed Lecture Notes), Princeton University, October 1935–January 1936, (cited as M.L.).

"An Unsolvable Problem of Elementary Number Theory," *American Journal of Mathematics*, LVIII, No. 2 (1936), pp. 345–63, (cited as U.P.).

(With J. B. Rosser) "Some Properties of Conversion," *Transactions of the American Mathematical Society*, Vol. 39, No. 3 (1936), pp. 472–82.

The Calculi of Lambda-Conversion, 1941, (cited as L.C.).

Church's work is an additional step in the process of generalizing logic. Actually it is a theory about functions such as are used in the various functional calculi. As he tells us in F.L. 1, the new system is inspired by three goals: (1) to avoid the use of the free variable, since extra-logical explanations are needed to clarify the meaning of the formulae in which they occur, (2) to avoid the confusions that arise when free and bound variables are used together, and (3) to avoid the

[1] A bibliography of Church's work will be found in *Journal of Symbolic Logic*, Vol. 1, No. 4 (1936), pp. 121–218. The system is the work of Church, Kleene, and Rosser.

paradoxes which were discussed earlier.[2] The notion, for example, of the operation of substitution is independent of the realm in which the substitution is made. Similarly, it may be possible to define abbreviation, regardless of what is to be abbreviated. In order to do this the general notion of "function" must be clarified and substitution allowed which will make the result of a substitution true, false, or meaningless. In this respect, Church approaches the intuitionist logic, although the two systems are not to be viewed as identical. Each modifies a different part of the principle of excluded middle.[3] Actually, as we shall see, Brouwer's rejection involves Church's but not conversely, since Brouwer rejects the principle of double negation for certain cases, while Church does not.

What Church is giving us is not logic but a meta-logic. The entities which are to be substituted for the variables in this system are the symbols of logic and mathematics. This, then, is a theory *about* logic and not a theory *of* logic. Its symbols are likewise generalizations but they refer to the properties of groups of symbols used in other systems. We must not confuse a metaphysical interpretation of logic with a metaphysical interpretation of such metalogics as this. Hilbert's notion of metamathematics is, in my opinion, not quite the same. Metamathematics is a theory of proof in mathematics, and that means it is really a logic. Mathematical entities are but one field of application of metamathematics, i.e., logic. But the fields of application of metalogics are symbols and nothing else. In $\lambda x[M](N)$ we can substitute only symbols. These latter symbols may refer to actual entities, but the symbols in $\lambda x[M](N)$ never do. In other words, Church's metalogic describes the characteristics of symbols which are used to refer to objects and their arrangement in the "sentences" of the object-language.

A "*function*" is defined as an operation capable of being applied to given things to obtain new things. Operation is taken as a primitive term. Thus in $+(2,3)$ the $+$ denotes a function acting on 2 and 3 to get 5. If the function is applied to one thing at a time it is a *function of one variable*. Thus if the function is $2(\ ) + 5$, where in the parentheses only one entity can be placed, we have a function of one variable. The object to which the function is applicable will be called *a value of the independent variable*. The class of such objects will be called the *range* of the independent variable. The object generated by the application of the function to a value of the independent variable is called a value of the dependent variable. $f(x)$ will denote the result of operating on x with the function f. Thus in mathematics $f(x) = x^2 + 2x + 1$ indicates a function of x, i.e., an operation on x which will result in new entities.

---

[2] F.L. 1, pp. 345–48.
[3] *Ibid.*, p. 348.

Let I be the function that associates any x with itself. Then $I(x) = x$ holds for any value of x whatsoever. I is called the *Identity Function*. The range of values x is the class of all objects, i.e., the universal class. There being no entities to which I is not applicable, the class of entities to which I is not applicable is empty and is called the *null class*.

Consider an ordinary function, say $3x^2 + 5x + 1$. If x is fixed and the calculation performed, this denotes a definite number. The result depends upon the operation of verifying x to be real and calculating as directed. We distinguish between the result of the operation and the operation itself. Consider a geometric illustration: If we place one leg of a compass at a given point and with a given separation of the legs of the compass rotate the other leg about the first we get a circle. Now we distinguish the operation with the compass from the circle. Returning, we shall denote the operation (i.e., the function) of verifying that x is an actual number and calculating the result of $3x^2 + 5x + 1$ by $\lambda x[3x^2 + 5x + 1]$.[4] This is to be distinguished from the result itself. Thus $\lambda x[3x^2 + 5x + 1] = \lambda y[3y^2 + 5y + 1]$ denotes that the two *operations* are identical. No value of x and y can be assigned here.

Suppose we have given the function $f(x) = 3x^2 + 5x + 1$ and also that x is 1. Then the operation of noting that 1 is a real number and inserting 1 in $3x^2 + 5x + 1$ will result in a value 9. We can then say $f(x) = 3x^2 + 5x + 1 = 9$. But if we wish to indicate the process itself which gives the result we write $\lambda x[3x^2 + 5x + 1]$ which now denotes "the operation of substituting a constant for x in $3x^2 + 5x + 1$ after ascertaining that the constant is suitable for that substitution." λx is called by Church "the functional abstraction operator." To say $3x^2 + 5x + 1 = 3y^2 + 5y + 1$ means that whenever the same substitutions are made for x as for y, the resulting values will be the same.

Another example is $\lambda x[x$ is a lioness$] = \lambda y[y$ is a daughter of a lioness$]$. Thus the λ operator is similar to the $\hat{x}$ of Russell.

When we have a variable x occurring as it does in λx where we cannot replace it by a constant we call it a *bound* variable. Such variables occur in mathematics in differentiation. For example, if $D_x(*)$ denotes the derivative of * with respect to x, then $D_x(3x^2 + 5x + 1)$ denotes a function (i.e., operation). It is not possible to substitute, say, 2 for x. What is ordinarily meant by $D_x(3x^2 + 5x + 1) = 6x + 5$ is $D(\lambda x[3x^2 + 5x + 1]) = \lambda x[6x + 5]$. Ordinary mathematical notation is thus ambiguous. If $f \equiv \lambda x[3x^2 + 5x + 1]$ then the value of this function for $x = 2$ will be denoted by $f(2) \equiv \{\lambda x[3x^2 + 5x + 1]\}(2)$. The λ here, and as the D_x of calculus, are some of the abstractions A_m referred to by von Neumann.

---

[4] In general x must be such an entity as to give meaning to the expression following λx.

A function of two variables will be defined as a function of one variable each of whose values is a function of one variable.

For example,

$$\{\{\lambda x[\lambda y[x^2 - 3xy - y^3]]\}(2)\}(3) = \{\lambda y[4 - 6y - y^3]\}(3) = [4 - 18 - 27] = -41.$$

Some of the braces, brackets, etc., may be omitted according to the following directions:

We can write $\lambda x{**}$ for $\lambda x[*,*]$
$\lambda xy{**}$ for $\lambda x[\lambda y[**]]$, and
$f(x)$ for $\{f\}(x)$.

Furthermore

1) the extent of an omitted square brackets shall be the shortest possible;
2) where the omission of a square bracket introduces an ambiguity a dot may be used instead;
3) when the dot appears explicitly it has the maximum possible scope, otherwise the minimum possible scope.[5]

We agree to use (for the present) the following symbols:

1. λ, [,], {,}, (,).
2. A countably infinite list of symbols $a, b, \ldots, x, y, \ldots$

Any symbol belonging to this list is called a *variable*. Any finite sequence of symbols will be called a *formula*. A symbol for a formula will be in boldface type.

Church now defines *inductively* the concepts "well-formed formula," "free occurrence of a variable," "bound occurrence of a variable." The well-formed formula is the "meaningful" formula. We noted earlier that the definition of "formula" allowed for any combination of symbols, e.g., [ ]λ{ }. In order to restrict the formulas to certain types that are ordinarily accepted in a system, the concept of well-formed is introduced.

I. Given a formula **A** with one symbol. **A** is well-formed if and *only* if the symbol is a variable. If a variable occurs in a well-formed formula with one symbol, its occurrence is a "free" occurrence.

II. Let n be a positive integer, we assume the meaning of well-formed to be understood for formulas involving symbols $(1 \leqq k \leqq n)$. Let **A** involve $n + 1$ symbols. **A** is *well-formed* (a) if it has the form

[5] F.L. 1, p. 354. Also M.L., p. 4.

$\{\mathbf{M}\}(\mathbf{N})$ where **M** and **N** are well-formed, or (b) if it has the form $\lambda\mathbf{x}[\mathbf{M}]$ and **M** is well-formed and x is a variable and at least one of occurrences of **x** in **M** is "free."[6]

If **A** is well-formed and is of form $\{\mathbf{M}\}(\mathbf{N})$, then an occurrence of a variable is *free* or *bound* in **A** according as it is free or bound in **M** or **N**.

If **A** is well-formed and of the form $\lambda\mathbf{x}[\mathbf{M}]$ then each occurrence of **x** is *bound* in **A** and each occurrence of a different variable **y** is *free* or *bound* in **A** according as it is free or bound in **M**.

For example, λxx is well-formed; $\{\lambda fx.f(x)\}(\mathbf{F}) = \lambda x\mathbf{F}(x)$ is well-formed if **F** is.

In what follows, "formula" always means "well-formed formula," boldface symbols represent such formulae, and "*wf*" denotes "well-formed" (This eliminates such "meaningless" formulae as [({()})]).

Let $S^{\mathbf{x}}_{\mathbf{A}}\mathbf{M}|$ denote the result of substituting **A** for **x** throughout **M**. For example, if **M** represents $3x^2 + 5x + 1$ and **A** the expression $y + 1$, then $S^{\mathbf{x}}_{\mathbf{A}}\mathbf{M}|$ would denote $3(y + 1)^2 + 5(y + 1) + 1$.

Operations: 0I. If the variable **y** is absent and every occurrence of the variable **x** is bound in a part **M** of **A**, then any particular occurrence of **M** in **A** may be replaced by $S^{\mathbf{x}}_{\mathbf{y}}\mathbf{M}|$.

Suppose we have **A** representing the expression $\lambda x.x(\lambda x.f(x,x))$. Consider $\lambda x.f(x,x)$ to be **M**. Then $S^{\mathbf{x}}_{\mathbf{y}}\mathbf{M}|$ would be $\lambda y.f(y,y)$ and **A** would become $\lambda x.x(\lambda y.f(y,y))$.[7] This enables us to change variables in a part of a formula.

0II. If **A** has a *wf* part of the form $\{\lambda\mathbf{x}[\mathbf{M}]\}(\mathbf{N})$ in which the variables occurring bound in **M** are distinct both from the variable **x** and from the free variables in **N**, then that part may be replaced by $S^{\mathbf{x}}_{\mathbf{N}}\mathbf{M}|$.

For example, if **A** is $\{\lambda x.x(x,y)\}(y(y))$ it has a *wf* part of the form $\{\lambda\mathbf{x}[\mathbf{M}]\}(\mathbf{N})$; then, since **N** is $y(y)$, we get $y(y)(y(y),y)$, which may be written $y(y)(y(y),y)$. This enables us to substitute one *wf* formula for the variable in another.

0III. If the variables occurring bound in **M** are distinct both from

[6] M.L., p. 5; also U.R., p. 346.

[7] This illustration is Church's in M.L., p. 6. Cf. also the statement of these rules of operation in U.P., p. 347, and the rules for substitution in Introduction to Mathematical Logic, Part 1, 1944, pp. 56–58 [Annals of Mathematics Studies, No. 13]. (This is not M.L.)

the variable **x** and from the free variable in **N** a *wf* part $S^{\mathbf{x}}_{\mathbf{N}}\mathbf{M}|$ of **A** may be replaced by $\{\lambda\mathbf{x}[\mathbf{M}]\}(\mathbf{N})$. This enables us to reverse the process described in 0II.

We define a *conversion*[8] as the operation of applying successively a finite sequence of operations 0I, 0II or 0III.

Aconv**B** means "there is a conversion applicable to **A** which gives **B**." The following properties are obvious:

Aconv**A**; **A**conv**B** implies **B**conv**A**; **A**conv**B** and **B**conv**C** implies **A**conv**C**. *Conv* is a symmetrical, reciprocal, and transitive operation. An example of conversion is $W(m,n)$conv$m + n$, where m,n represent positive integers; also $[m + n] + t$conv$m + [n + t]$.

We define a *reduction* as a conversion in which 0III is not used and 0II occurs once and only once.

Ared**B** means there is a finite sequence of reductions applicable to **A** which gives **B**. **A**red**B** is transitive but not symmetric.

A is said to be in *normal form* if it contains no part of the form $\{\lambda x[\mathbf{M}]\}(\mathbf{N})$ where **M** and **N** are *wf*.

This means that **A** contains no part to which 0II is applicable. For example, $\lambda ac.c(a)$ is in normal form. It follows as an immediate theorem from the definition of normal form and reduction that if a formula is in normal form no reduction of it is possible. It is important to note that a formula is *meaningful* if, and *only* if, it has a normal form.[9]

B is a *normal form of* **A** if **B** is in normal form and **A**conv**B**. An example of a formula that has no normal form is $x.x(x)(x,x(x))$, since 0II can always be applied to it, it being invariant under applications of 0II.

a → **A** shall denote that the particular symbol **a** shall represent some particular sequence of symbols **A**. **A** will ordinarily be *wf* containing no free variables. This arrow may be read "stands for" or "is an abbreviation of." The usual logical operations can be defined[10] as

$V \rightarrow \lambda\mu\nu.\sim.\sim\mu.\sim\nu$ ("either-or")

$U \rightarrow \lambda\mu\nu.\sim.\mu.\sim\nu$ ("implies")

$Q \rightarrow \lambda\mu\nu.\Pi(\mu,\nu).\Pi(\nu,\mu)$ ("equivalent")

$E \rightarrow \lambda\Pi\Sigma\phi.\phi(\Pi)$ ("exists")

For example:

$V(\mathbf{p},\mathbf{q}) \rightarrow \lambda\mathbf{pq}.\sim.\sim\mathbf{p}.\sim\mathbf{q}$ which reads "**p** or **q**."

Note that the formulae are the abstract frameworks of logical forms.

---

[8] Cf. for properties of conversion: Church and Rosser, "Some Properties of Conversion," *Trans. Amer. Math. Soc.*, Vol. 39 (1936), pp. 472–82.

[9] Cf. U.P., p. 348; also M.L., pp. 7 and 17.

[10] F.L. 1, p. 355.

We introduce:

$$I \rightarrow \lambda xx$$
$$1 \rightarrow \lambda fx.f(x)$$
$$2 \rightarrow \lambda fx.f(f(x))$$
$$3 \rightarrow \lambda fx.f(f(f(x)))$$
$$4 \rightarrow \lambda fx.f(f(f(f(x)))) \text{ etc.}$$

The formulae 1, 2, 3, 4,... represent the positive integers. This sequence has the same internal structure as the sequences of positive integers.

We can obtain the successor $\mathbf{k} + \mathbf{1}$ to any integer $\mathbf{k}$ by means of $S \rightarrow \lambda nfx.f(n(f,x))$ and the property $S(\mathbf{k})\text{conv}\mathbf{k} + \mathbf{1}$.

We introduce the following abbreviations, which are needed to define certain operations and theorems about numbers:

Let $B \rightarrow \lambda fgx.f(g(x))$

B stands for the result obtained by operating on an operation, i.e., it represents a function of a function. It is also viewed as an associative operator.

$$C \rightarrow \lambda fxy.f(y,x)$$

C is the operation of interchanging two variables. If they are interchanged in a function of them which is equated to 0, it becomes the operation of taking the converse of the function of one of the variables which was first defined implicitly by the equation.

$$W \rightarrow \lambda fx.f(x,x)$$

W is the operator which allows a repetition of a variable. For example:

$$Wxy = xyy$$
$$[\mathbf{m},\mathbf{n}] \rightarrow \lambda fgx.\mathbf{m}(f,\mathbf{n}(g,x))$$
$$[\mathbf{m},\mathbf{n},\mathbf{t}] \rightarrow \lambda fghx.\mathbf{m}(f,\mathbf{n}(g,\mathbf{t}(h,x))).$$

If we let

$$H \rightarrow \lambda m.[S(\lambda f.m(f,I,I)),\lambda f.m(f,I,I),\lambda f.m(I,f,I)]$$

then we can define the predecessor operator as:

$$P \rightarrow \lambda n.n(H,[I,I,I],I,I).$$

This selects the predecessor to any number.[11] Since we are dealing with positive integers, the predecessor of 1 is taken as 1.

---

[11] For additional comments on some of these operators, refer to L.C., pp. 4–5, and to p. 31 for the P operator. M.L., p. 21, proves that if n represents a positive integer greater than 1, $P(n)\text{conv}n - 1$. Cf. also H. B. Curry, "Grundlagen der kombinatorischen Logik, Teil I," *Amer. Jour. Math.*, Vol. 52 (1930), pp. 509 ff.; Teil II, *ibid.*, pp. 789 ff. Also see H. B. Curry, "The Combinatory Foundations of Mathematical Logic," *Journal of Symbolic Logic*, Vol. 7 (1942), pp. 49 ff.

$$[\mathbf{m}] + [\mathbf{n}] \rightarrow \lambda fx . \mathbf{m}(f,\mathbf{n}(I,x)).$$

$$[\mathbf{m}] \times [\mathbf{n}] \rightarrow B(\mathbf{m},\mathbf{n}).$$

(For the application to the calculus of propositional functions the reader is referred to M.L., Chapter 2, pp. 14 ff., and to L.C.)

To continue with the development, note that the positive integers are viewed as abbreviations for the terms of a sequence of λ-formulae having the same internal structure as the sequence of natural numbers. In other words, the 2 defined as above represents the structure of the definition of the cardinal number 2.

Let **f** be a formula and $\mathbf{m}_1,\mathbf{m}_2,\mathbf{m}_3,\ldots\mathbf{m}_r$ be formulae representing positive integers. Let f be a function of r positive integers $m_1,m_2,m_3,\ldots m_r$ corresponding to the function **f** of the formulas $\mathbf{m}_1,\mathbf{m}_2,\ldots,\mathbf{m}_r$ to which $m_1,m_2,\ldots,m_r$ correspond. The function f is said to be *lambda-definable* if we can find a formula **f** such that

$$\{\mathbf{f}\}(\mathbf{m}_1,\mathbf{m}_2,\ldots,\mathbf{m}_r)\text{conv}\{f\}(m_1,m_2,\ldots,m_r).$$

We now prove that addition and multiplication are lambda-definable. That is, we want to prove that if **m** and **n** are formulae representing positive integers m and n, then $[\mathbf{m}]+[\mathbf{n}]\text{conv}\mathbf{m}+\mathbf{n}$, $[\mathbf{m}\times\mathbf{n}]\text{conv}\mathbf{m}\times\mathbf{n}$.

For $[\mathbf{1}]+[\mathbf{n}]\text{conv}S(\mathbf{n})\text{conv}\mathbf{1}+\mathbf{n}$.

Suppose it true for $m=k$ (i.e., $[\mathbf{k}]+[\mathbf{n}]\text{conv}\mathbf{k}+\mathbf{n}$) and consider $m=k+1$.

$[\mathbf{k}+\mathbf{1}]+[\mathbf{n}]\text{conv}[S(\mathbf{k})]+[\mathbf{n}]\text{conv}[\lambda fx.f(\mathbf{k}(f,x))]+[\mathbf{n}]\text{conv}\lambda fx.\{\lambda y.f(\mathbf{k}(f,y))\}(\mathbf{n}(f,x))\text{conv}\lambda fx.f(\mathbf{k}(\mathbf{f},\mathbf{n}(fx)))\text{conv}\lambda fx.f(\{[\mathbf{k}]+[\mathbf{n}]\}(f,x))$ $\text{conv}S([\mathbf{k}]+[\mathbf{n}])\text{conv}S(\mathbf{k}+\mathbf{n})\text{conv}\mathbf{k}+\mathbf{1}+\mathbf{n}$.

Hence, by induction for all m, $[\mathbf{m}]+[\mathbf{n}]\text{conv}\mathbf{m}+\mathbf{n}$.

For the proof of $[\mathbf{m}]\times[\mathbf{n}]\text{conv}\mathbf{m}\times\mathbf{n}$ we first note that $1(\lambda\mathbf{x}\mathbf{M})$ $\text{conv}\lambda\mathbf{x}\mathbf{M}$.

Now $[\mathbf{1}]\times[\mathbf{n}]\text{conv}\lambda x.1(\mathbf{n}(x))\text{conv}\lambda x\mathbf{n}(x)\text{conv}\mathbf{n}$.

Suppose when $m=k$, $[\mathbf{k}]\times[\mathbf{n}]\text{conv}\mathbf{k}\times\mathbf{n}$. Consider $m=k+1$, $[\mathbf{k}+\mathbf{1}]\times$ $\mathbf{n}\text{conv}[S(\mathbf{k})]\times\mathbf{n}\text{conv}\lambda x.S(\mathbf{k},\mathbf{n}(x))\text{conv}\lambda xy\mathbf{n}(x,\mathbf{K}(\mathbf{n}(x),y))\text{conv}\lambda xy.\mathbf{n}$ $(x\{[\mathbf{k}]\times[\mathbf{n}]\}(x,y)\text{conv}[\mathbf{n}]+[[\mathbf{k}]\times[\mathbf{n}]]\text{conv}[\mathbf{n}]+[\mathbf{k}\times\mathbf{n}]\text{conv}\mathbf{n}+(\mathbf{k}\times$ $\mathbf{n})\text{conv}(\mathbf{k}+\mathbf{1})\times\mathbf{n}$.

Hence, by induction for all m,$[\mathbf{m}]\times[\mathbf{n}]$conv$\mathbf{m}\times\mathbf{n}$. We define also:

$$[\mathbf{m}]-[\mathbf{n}]\rightarrow \mathbf{n}(P,\mathbf{m})$$

m and **n** stand for formulae representing positive integers. As a consequence $[\mathbf{m}]-[\mathbf{n}]$ is defined only for such integers. For all cases where $[\mathbf{m}]$ represents a positive integer greater than the positive integer represented by $[\mathbf{n}]$, the definition gives us the subtraction of ordinary arithmetic, for $[\mathbf{m}]-[\mathbf{n}]$conv$\mathbf{m}-\mathbf{n}$. Where $[\mathbf{m}]$ represents an integer less than or equal to that represented by $[\mathbf{n}]$, $[\mathbf{m}]-[\mathbf{n}]$conv1.

$$\text{Min}\rightarrow \lambda xy.S(y)-.S(y)-x$$

$\text{Max}\rightarrow \lambda xy.[x+y]-\min(x,y)$. Min$(\mathbf{m},\mathbf{n})$ represents the lesser of m and n; max the greater.

We use the formula 2 to represent the truth-value T and 1 to represent F. Then a function of positive integers for which the values of the dependent variable are 2 and 1 may be regarded as a propositional function of positive integers.

The propositional functions "is greater than" and "equals" for positive integers are lambda-definable as follows:

$$e\rightarrow \lambda xy.\min(2,S(x)-y).$$
$$d\rightarrow \lambda xy.4-e(x,y)+e(y,x).$$

If **m** and **n** represent positive integers, $e(\mathbf{mn})$conv2, if $m>n$, otherwise $e(\mathbf{m},\mathbf{n})$conv1; (since 2 is used to represent the truth value T, $e(\mathbf{m},\mathbf{n})$conv2 if $m>n$ means $e(m,n)$ is true). If $m=n$, $d(\mathbf{m},\mathbf{n})$conv2, otherwise $d(\mathbf{m},\mathbf{n})$conv1.

Let $I\rightarrow \lambda xx$
$J\rightarrow \lambda fxyz.f(x,f(z,y))$

We define a *combination* as follows:

I or J or any variable standing alone is a combination. If **A** and **B** are combinations, $\{\mathbf{A}\}(\mathbf{B})$ is.

A combination is therefore any *wf* formula obtainable from I and J and variables by joining them together in braces and parentheses in any permutation.

It can be proved now that every formula can be converted into a combination:[12]

We define a combination belonging to a *wf* formula as

(1) the formula itself, if the formula is of one symbol.

[12] For the proof refer to Church, *Mathematical Logic*, Princeton Univ. Press (1935–36), pp. 24 ff. It is too long to reproduce.

(2) as $\{\mathbf{M}'\}(\mathbf{N}')$ if the formula is of more than one symbol and has the form $\{\mathbf{M}\}(\mathbf{N})$ and $\mathbf{M}'$ and $\mathbf{N}'$ are the combinations belonging to **M** and **N** respectively.

(3) as $\lambda\mathbf{x}[\mathbf{R}']|$ if the formula is of more than one symbol and has the form $\lambda\mathbf{x}[\mathbf{R}]$ and $\mathbf{R}'$ is the combination belonging to **R**.

It is clear, says Church, that there is a unique combination to any given formula and a constructive method of obtaining this combination. There is a constructive method also for ascertaining whether there is a formula to which a given combination belongs and of obtaining such a formula if there is one. It is possible to re-define some of the concepts previously used in terms of combinations.[13] Given the I and J as defined above and let $T \rightarrow J(I,I)$

then $C \rightarrow J(T,J(T),J(T))$

$B \rightarrow C(J(I,C),J(I))$

$[\mathbf{p}]\times[\mathbf{q}] \rightarrow B(\mathbf{p},\mathbf{q})$

$1 \rightarrow B(I)$

$\phi \rightarrow B(B,W)\times B(C)\times B(B,B)\times B$

$S \rightarrow \phi(B,1)$

$2 \rightarrow S(1)$

$3 \rightarrow S(2)$

Any function of positive integers F can always be associated with an infinite sequence $F(1)$, $F(2)$, ... We can therefore speak of a formula **F** for a function F of positive integers as representing an infinite sequence and write either **F**repr**F**(1), **F**(2)... or **F**repr$F(1)$, $F(2)$,....

A function is called a *function of positive integers* if the range of each independent variable is the class of positive integers and the range of the dependent variable is contained in the class of positive integers.[14]

A new symbol δ (standing for the notion of equality as a propositional function of two variables) is now introduced and certain of the original concepts are redefined. This extends the system so as to make possible the derivation of a system of symbolic logic as well as the concept of positive number. What is done is to add to the definitions and rules of the system additional parts about the new symbol. The δ is introduced to make possible an interpretation of the calculus in terms of extension, and to develop a system of symbolic logic which is ultimately finitistic in character.

A formula of one symbol shall be called *well-formed* if it consists of a

[13] M.L., p. 30.
[14] U.P., p. 348.

variable or of δ (δ is not a variable). For more than one symbol—the definition follows the one given above.

Instead of *normal-form* we have two concepts replacing the one:

1) *lambda-normal form* which has the definition given "normal form" above;

2) *delta-normal form*, possessed by all (and *only* those) *wf* formulas which have neither of the two properties:

a) having a *wf* part of form $\{\lambda x\mathbf{M}\}(\mathbf{N})$
b) having a *wf* part of form $\delta(\mathbf{M},\mathbf{N})$ which contains no free variables.

(We shall use *normal form* to mean *delta-normal form*.)

Two new operations, which leave the intuitive meaning of the formulae to which they are applied invariant, are introduced:

0IV. If a formula contains a *wf* part $\delta(\mathbf{M},\mathbf{N})$ in which **M** and **N** are in delta-normal form and contain no free variables, then $\delta(\mathbf{M},\mathbf{N})$ may be replaced by 2 if **M** and **N** are the same to within 0I or by 1 if **M** and **N** are not the same to within 0I (i.e., any part $\delta(\mathbf{M},\mathbf{N})$ may be replaced by 1, under the given conditions, provided also that **M** is not convertible into **N** by 0I. This condition is not necessary in replacing $\delta(\mathbf{M},\mathbf{N})$ by 2).

0V. If a formula contains a part 2, that part may be replaced by $\delta(\mathbf{M},\mathbf{N})$ where **M** and **N** are in delta-normal form, contain no free variables, and are the same to within 0I. If a formula contains a part 1, that part may be replaced by $\delta(\mathbf{M},\mathbf{N})$; **M** and **N** in delta-normal form, contain no free variables and are not the same to within 0I.

0IV and 0V are such that they can always be carried out effectively.[15]

A finite sequence of these operations 0I–0V is called a *conversion*.[16] If the conversion uses only 0I–0III it is called a *lambda-conversion*. If the conversion contains one application of 0II or 0IV and all its steps besides this are applications of 0I, it is called a *reduction*.

Ordered pairs can now be introduced in general:

We define $[\mathbf{a},\mathbf{b}] \rightarrow \lambda f.f(\mathbf{a},\mathbf{b})$ where **a** and **b** are any *wf* formulae. (This is the general form of which the definition $[\mathbf{m},\mathbf{n}]$ above is a special case. Here the **a** and **b** represent any *wf* formulas, while **m** and **n** represent formulas corresponding to numbers.)

---

[15] For additional remarks on conversion, *see* U.P., pp. 347 and 358. Cf. also the more precise statement in L.C., pp. 62–63.

[16] Cf. also Church and Rosser, "Some properties of Conversion," *Trans. Amer. Math. Soc.*, Vol. 39 (May, 1936), pp. 472–82.

A constancy operator **K**, the purpose of which is to change a constant into a function whose value is the constant is defined.

$K \rightarrow \lambda xy.\delta(y,y,I,x)$. The way in which δ operates can be shown by this definition. Given $\delta(y,y.I,x)$. The δ operates on the y,y to reduce $\delta(y,y)$ to 2; the 2 operates on I to reduce $2(I)$ to I, and the $I(x)$ reduces to x—provided y is replaced by any *wff* with no free variables which has a normal form, then

$\{[\mathbf{a},\mathbf{b}]\}(K)\text{conv}K(\mathbf{a},\mathbf{b})\text{conv}\delta(\mathbf{b},\mathbf{b},I,\mathbf{a})$ and if **b** has a delta-normal form with no free variables,

$$\{[\mathbf{a},\mathbf{b}]\}(K)\text{conv}\mathbf{a}$$

also $\{[\mathbf{a},\mathbf{b}]\}(K(I))\text{conv}K(I,\mathbf{a},\mathbf{b})\text{conv}\delta(\mathbf{a},\mathbf{a},I,I,\mathbf{b})$ and if **a** has a delta-normal form with no free variables

$$\{[\mathbf{a},\mathbf{b}]\}(K(I))\text{conv}\mathbf{b}.$$

Another concept is now introduced—that of *metads*[17]—which will enable us to develop a definition of the concept *positive integer*. These symbols will also enable us to make statements about the formula itself, whether or not it has any meaning.[18] These metads offer modes of expression about symbols and not about the meanings of the symbols. As we shall see, the expression "to be a natural number" will be defined in terms of metads. As a consequence this expression is about the *symbols* which are used to represent natural numbers. It has nothing to do with the *meaning* of "natural number" except very indirectly.

The purpose of the metad is to make it possible to translate statements about formulae into arithmetic statements. In this way this whole system is transformed into a system of arithmetic. A somewhat similar process will be described later, one which was constructed by Gödel and which led to the theorem that the proposition: "the system S (which satisfies conditions to be described later) is consistent" is an undecidable proposition within S.

A metad is a symbol that stands for a formula considered as an entity. Metads will have normal forms, i.e., meanings, whether or not the formulae they represent have normal forms. The meaning of a metad will, of course, be the formula it represents. These symbols are called metads because they are symbols about symbols. It will be recalled that a positive whole number representing, as it does, a property of a class which is itself an abstraction is a metad.

---

[17] The discussion follows M.L., pp. 51 ff. Cf. also L.C., p. 68.

[18] Church has since given another approach to this problem through Gödel numbers. Cf. L.C., pp. 51 ff. The present approach is retained as illustrative of Church's techniques and because Church refers to "met" in L.C.

Definition of *metads*:

1. The formulae which stand for the positive integers are metads of rank 1.

2. If **a** and **b** are metads, [**a**,**b**] is a metad whose rank is greater by one than the greater of the two ranks of **a** and **b**.

Definition: Metad representing a combination.

1. The metad of I is 1.
   The metad of J is 2.
   The metad of δ is 3.

   The metads of the countable infinite set of variables $a,b,\ldots,$ $x,y,\ldots$ are in order of the enumeration 4, 5, 6,...

2. If **a** is the metad of **p**, and **b** the metad of **q** where **p** and **q** are combinations, then [**a**,**b**] is the metad of **p**(**q**).

Definition: the metad belonging to a formula is the metad which represents the combination belonging to the formula.

It can easily be proved that every metad has a normal form.

Definition: *Terms* of a metad.

1. The metads of rank one have only one term, which is the metad itself.

2. The terms of [**a**,**b**] are, in order, the terms of **a** followed by the terms of **b**.

Definition: *Primary* metad.

1. 1, 2, and 3 are primary metads.

2. If **a** and **b** are primary metads, [**a**,**b**] is. A primary metad then is one whose terms consist entirely of 1's, 2's, and 3's.

Definition: A *normal* metad is a metad belonging to a formula in normal form.

We define *form* and *met*—two functions—making possible within the symbolism, the passage from a formula to its metad and from the metad to the formula; i.e., form taken of a metad gives the formula to which it belongs; met taken of a formula gives the metad belonging to the normal form of that formula; e.g., form(**a**)conv**A** if **a** belongs to **A** and **A** contains no free variables. Met(**A**)conv**a** belongs to **A**′ the normal form of **A**. These functions enable us to pass from any formula to the symbol which represents it, and conversely.

Let $Y \rightarrow \lambda f.\sim\delta(1,f(\lambda xy.\delta(x,x,\delta(y,y,1))))$. It can be proved that if **a** is a metad of rank 1, $\Upsilon(a)$conv2; if **a** is a metad of rank greater than 1, $\Upsilon(a)$conv1.

Two functions are now defined which will give Υ what shall be called "the first member" and "the second member" of any metad.

We find an m such that

$m_1(1)\text{conv}\lambda f.f(K)$ (K is the constancy function)
$m_1(2)\text{conv}\lambda f.f(1)$

and an m_2 such that

$m_2(1)\text{conv}\lambda f.f(K(I))$.
$m_2(2)\text{conv}I$.

Then define $M_1 \rightarrow \lambda s.m_1(\Upsilon(s),s)$

$M_2 \rightarrow \lambda s.m_2(\Upsilon(s),s)$

M_1 is read "first member of"; M_2 "second member of."

We abbreviate $\mathbf{a}_1 \rightarrow M_1(\mathbf{a})$; $\mathbf{a}_2 \rightarrow M_2(\mathbf{a})$;$\mathbf{a}_{21} \rightarrow M_1(M_2(\mathbf{a}))$; $\mathbf{a}_{122} \rightarrow M_2(M_2(M_1(\mathbf{a})))$ etc.

The definition of the rank of any metad a involves the notion of "the least positive integer n not less than k satisfying some function of positive integers," since the rank is to be taken as a least number. It is necessary, therefore, to introduce a function whose properties enable it to select such a number. Church uses the Kleene p-function defined as follows:

$p \rightarrow \lambda yk.H(y(k),G,y,k)$

where H and G are formulas such that

$H(1)\text{conv}\lambda xyk.x(1,y(S(k)),x,y,S(k))$
$H(2)\text{conv}\lambda xyk.x(2,y(k,I),k)$
$G(1)\text{conv}H$
$G(2)\text{conv}I$

Now, if D represents a propositional function of positive integers, Church shows that $p(D,k)$ is "convertible into the formula n which represents the least positive integer n not less than k, for which $D(n)$ conv2" or, if no such n exists, then $p(D,k)$ has no normal form.

Let $\theta \rightarrow \lambda fx.f(x_1).f(x_2)$ and

$|\mathbf{a}| \rightarrow p(\lambda n.n(\theta,\Upsilon,[\mathbf{a},\mathbf{a}])1)$: read, "rank of a."

To define *form*, we find an H such that

$H(1)\text{conv}I$
$H(2)\text{conv}\mathcal{J}$
$H(3)\text{conv}\delta$

and then a G such that

$G(1)\text{conv}H$
$G(S(\mathbf{n}))\text{conv}\lambda x.G(\mathbf{n},x_1,G(\mathbf{n},x_2))$ for any positive integer n. Then definition: form $\rightarrow \lambda x.G(|x|,x)$.

Now we need to define *met*. In order to do so, it is necessary to find first a formula 0 such that $0(\mathbf{a})\text{conv}2$ if $\mathbf{a}$ is a metad representing a combination of the form $\lambda_{\mathbf{x}}\mathbf{p}|$ and $0(\mathbf{a})\text{conv}1$ if $\mathbf{a}$ is a metad representing a combination not of this form. First we find a G such that:

$G(1)\text{conv}\delta(1)$
$G(2)\text{conv}\lambda x.\delta([[2,[[2,1],1]],[[2,1],1]],x_1).\delta([2,1],x_{21})$
$.\delta([[2,[[2,1],1]],[[2,1],1]],x_{221})$
$.\delta([2,[[2,1],1]],x_{22211}).\delta([2,[[2,1],1]],x_{222211}).\delta(2,x_{22222})$
$G(3)\text{conv}\lambda x.\delta([2,[[2,1],1]],x_{11}).\delta([2,1],x_{21})$
$G(4)\text{conv}\lambda x.\delta([2,[[2,1],1]],x_{11})$
$G(5)\text{conv}E.$

Now let $H \rightarrow \lambda x.p(\lambda n.G(n,x),1)$.

We then have a theorem:

Suppose $\mathbf{a}$ is a metad. Then

1. If $\mathbf{a}$ represents I, then $H(\mathbf{a})\text{conv}1$.

If not, then

2. If $\mathbf{a}$ represents a combination of the form

$\mathcal{J}(T,T,\mathcal{J}(I,\mathcal{J}(T,T,\mathcal{J}(T,\mathbf{A},\mathcal{J}(T,\mathbf{F},\mathcal{J})))))$

then $H(\mathbf{a})\text{conv}2$.

If not, and

3. If $\mathbf{a}$ represents a combination of the form $\mathcal{J}(T,\mathbf{A},\mathcal{J}(I,\mathbf{F}))$,

then $H(\mathbf{a})\text{conv}3$.

If not, and

4. If $\mathbf{a}$ represents a combination of the form $\mathcal{J}(T,\mathbf{A},\mathbf{F})$

then $H(\mathbf{a})\text{conv}4$.

5. If not, $H(\mathbf{a})$conv5.

We shall omit the proof.[19]

We next find an n_1 such that

$n_1(1)\text{conv}K(1)$
$n_1(2)\text{conv}\lambda x\,.\,x_{222212}$
$n_1(3)\text{conv}K(1)$
$n_1(4)\text{conv}\lambda x\,.\,x_2$
$n_1(5)\text{conv}K(2)$

and an n_2 such that

$n_2(1)\text{conv}K(1)$
$n_2(2)\text{conv}\lambda x\,.\,x_{22212}$
$n_2(3)\text{conv}\lambda x\,.\,x_{12}$
$n_2(4)\text{conv}K(1)$
$n_2(5)\text{conv}K(2)$

now let $\mathcal{N}_1 \rightarrow \lambda x\,.\,n_1(H(x),x)$
$\mathcal{N}_2 \rightarrow \lambda x\,.\,n_2(H(x),x)$
$R \rightarrow \lambda fx\,.f(\mathcal{N}_1(x))\,.f(\mathcal{N}_2(x))$

Then $o \rightarrow \lambda x\,.\,\{|x|\}(R,\delta(1),x)$.

Having now this definition, we need a few more concepts before *met.*, and then *natural number*, can be defined.

We wish to define a function U such that, if $\mathbf{a}$ is a metad representing $\lambda_{\mathbf{x}}\mathbf{p}|$ and if $\mathbf{y}$ is the first variable in the list $a, b, \ldots x, y \ldots$ which does not appear free in $\lambda_{\mathbf{x}}\mathbf{p}|$, then $U(\mathbf{a})$ is a metad representing $S^{\mathbf{x}}_{\mathbf{y}}\mathbf{p}|$; and if $\mathbf{a}$ is a metad representing a combination not of this form, then $U(\mathbf{a})$ conv$\mathbf{a}$

$\lambda_{\mathbf{x}}\mathbf{M}|$, where $\mathbf{M}$ is any combination containing $\mathbf{x}$ as a free variable, is defined as follows:

$\lambda_{\mathbf{x}}x|$ is I

Three cases are possible in $\lambda_{\mathbf{x}}\mathbf{F}(\mathbf{A})|$

(1) $\mathbf{A}$ contains $\mathbf{x}$ as a free variable, but $\mathbf{F}$ does not; then $\lambda_{\mathbf{x}}\{\mathbf{F}\}(\mathbf{A})|$ is $\mathcal{J}(T,\lambda_{\mathbf{x}}\mathbf{A}|,\mathcal{J}(I,\mathbf{F}))$.

(2) $\mathbf{F}$ contains $\mathbf{x}$ as a free variable but $\mathbf{A}$ does not; then $\lambda_{\mathbf{x}}\{\mathbf{F}\}(\mathbf{A})$ is $\mathcal{J}(T,\mathbf{A},\lambda_{\mathbf{x}}\mathbf{F}|)$

[19] Church, *Math. Logic*, p. 58.

(3) **F** and **A** contain **x** as a free variable;

then $\lambda_{\mathbf{x}}\{\mathbf{F}\}(\mathbf{A})|$ is $\mathcal{J}(T,T,\mathcal{J}(I,\mathcal{J}(T,T,\mathcal{J}(T,\lambda_{\mathbf{x}}\mathbf{A}|,\mathcal{J}(T,\lambda_{\mathbf{x}}\mathbf{F}|,\mathcal{J})))))$ where I and $\mathcal{J}$ are as previously defined and $T \rightarrow \mathcal{J}(I,I)$.

Let $\gamma \rightarrow \lambda nx.\{|x|\}(\theta,\lambda y.\sim\delta(n.y),x)$ where, as before, $\theta \rightarrow \lambda fx.f(x_1).f(x_2)$. Then if **a** is a metad and **n** represents a positive integer greater than one, $\gamma(\mathbf{n},\mathbf{a})$conv1 represents a positive integer greater than one; $\gamma(\mathbf{n},\mathbf{a})$conv1 if **a** has **n** as one of its terms; otherwise $\gamma(\mathbf{n},\mathbf{a})$conv2.

Let $ary \rightarrow \lambda x.\{|x|\}(\theta,\lambda x.\delta(1,x) \vee \delta((3,x),x)$.

Find a G such that

$G(1)\text{conv}\lambda x.\sim o(x)$

$G(2)\text{conv}S(1)$

$G(3)\text{conv}\lambda x.\delta([[2,[[2,1],1]],[[2,1],1]],x_1)$
$.\delta([2,1],x_{12}).\delta([[2,[[2,1],1]],[[2,1],1]],x_{221})$
$.\delta([2,[[2,1],1]].x_{22211}).\delta([2,[[2,1],1]]x_{222211})$
$.\delta(2,x_{22222})$

$G(4)\text{conv}\lambda x.\delta([2,[[2,1],1]]x_{11}).\delta([2,1],x_{21})$

$G(5)\text{conv}\lambda x.\delta([2,[[2,1],1]],x_{11})$

Let $H \rightarrow \lambda x.p(\lambda n.G(n,x),1)$. Then again for this H, if **a** is a metad, $H(\mathbf{a})$conv1, if **a** represents a combination not of the form $\lambda_{\mathbf{x}}p|$. If not, then $H(\mathbf{a})$conv2 if **a** represents I. If not, $H(\mathbf{a})$conv3 if **a**conv3 if **a** represents a combination of the form

$$\mathcal{J}(T,T,\mathcal{J}(I,\mathcal{J}(T,T,\mathcal{J}(T,\mathbf{A},\mathcal{J}(T,\mathbf{F},\mathcal{J}))))).$$

If not, $H(\mathbf{a})$conv4 if **a** represents a combination of the form $\mathcal{J}(T,\mathbf{A},\mathcal{J}(I,\mathbf{F}))$. If not, $H(\mathbf{a})$conv5 since **a** represents a combination of the form $\mathcal{J}(T,\mathbf{A},\mathbf{F})$.

Now find an X such that

$X(1)\text{conv}\lambda fx.f(1,I,x)$

$X(2)\text{conv}\lambda fx.f(x)$

$X(3)\text{conv}\lambda fx.[f(x_{222212}),f(x_{22212})]$

$X(4)\text{conv}\lambda fx.[x_{22},f(x_{12})]$

$X(5)\text{conv}\lambda fx.[f(x_2),x_{12}]$

and let $Z \rightarrow \lambda fx.X(H(x),f,x)$

then $U \rightarrow \lambda x.\{|x|\}(Z,\lambda m.m(I,p(\lambda n.y(n,x)4)),x)$

We now wish to define a propositional function for the normality of a metad.

Find a G such that

$G(1)\text{conv}o$
$G(2)\text{conv}\lambda x.[o(x_1).\sim\gamma(x)] \vee \delta(3,x_{11})ary(x_{12})ary(x_2)$
$G(3)\text{conv}E$

Let $H \rightarrow \lambda x.p(\lambda n.G(n,x),1)$. Now find an n_1 and n_2 such that

$n_1(1)\text{conv}U$ $n_2(1)\text{conv}U$
$n_1(2)\text{conv}K(2)$ $n_2(2)\text{conv}K(2)$
$n_1(3)\text{conv}M_1$ $n_2(3)\text{conv}M_2$

Let $N_1 \rightarrow \lambda x.n_1(H(x),x)$ and $N_2 \rightarrow \lambda x.n_2(H(x),x)$.
$R \rightarrow \lambda fx.f(N_1(x)).f(N_2(x))$ and we now define
$al \rightarrow \lambda x\{|x|\}(R,\lambda n.\sim\delta(2,n),x)$.

We can find a formula *Prim* for which Prim (1), Prim (2), Prim (3) . . . is an enumeration (although with repetitions) of all primary metads.

Let $Prn \rightarrow \lambda n.\text{Prim}(P(\lambda x.al(\text{Prim}(x)),n))$. Then $Prn(1)$, $Prn(2)$, $Prn(3)$. . . is an enumeration of the primary normal metads.

Then $\text{met} \rightarrow \lambda x.Prn(\text{p}(\lambda n.\delta(\text{form}(Prn(n)),x),1))$.

We are now in a position to define the class of positive integers by means of a propositional function N, conv2 if applied to a formula representing a positive integer.

Construct a G such that

$G(1)\text{conv}K(2)$ $G(2)\text{conv}M_2$

Let $H \rightarrow \lambda x.G(\delta(4,x_1),x)$ and

$N' \rightarrow \lambda x.o(U(x)).\delta(5,P(|U(U(x))|,H,U(U(x))))$

If **a** is a metad then $N'(a)$conv2 if **a** belongs to a formula in normal form and represents a positive integer, otherwise $N'(\mathbf{a})$conv1.

That this is true follows from the fact that $N'(\mathbf{a})$conv2 if and only if **a** has the following two properties:

1) The formula to which **a** belongs begins with a λ, and still has a λ at the beginning after the first one is cut off.

2) After applying H to **a** as many times in succession as the predecessor of the rank of the metad which belongs to the formula with both λ's cut off at the beginning, we are left with 5.

From the way H operates on a metad, it follows that $U(U(\mathbf{a}))$ must represent $\underbrace{a(a(\ldots(a,(b))\ldots))}_{P(|U(U(\mathbf{a}))|)\text{ times}}$, where a and b are the first two variables in the list $a, b, c, \ldots x, y, \ldots$ Therefore, **a** belongs to a formula which may be written $\lambda fx.f(f(\ldots f(f(x))\ldots))$ which represents a positive integer. The P in this formula represents a function such that, if **D** represents a propositional function of positive integers, $P(D)$ represents the infinite sequence of positive integers $n_1 n_2 n_3 \ldots$ if $\mathbf{D}(\mathbf{n})$conv2 for n equal to $n_1, n_2, n_3, \ldots$

The P is defined:

$P \rightarrow \lambda yn.H(n,y)$, where H is such that
$H(1)\text{conv}\lambda y.p(y,1)$
$H(S(n))\text{conv}\lambda y.p(y,S(H(\mathbf{n},y)))$.

Finally, $\mathcal{N} \rightarrow \lambda x.\mathcal{N}'(\text{met}(x))$. $\mathcal{N}(\mathbf{M})$conv2 if **M** is a formula standing for a positive integer; otherwise, if **M** is a formula having a normal form, $\mathcal{N}(\mathbf{M})$conv1.

The extreme symbolization is quite obvious. Yet the insistence on *calculability* which runs through the entire analysis is a step towards a *rapprochement* with the intuitionists. Church also seems to insist on the idea of construction rather than mere consistency. The notion of λ-definability is extremely important and has applications in the discussion of consistency. It is, however, too advanced for treatment here.

Another important point in this analysis is the clearer distinction between an analysis of a language and an analysis of the objects to which the language is applied (in other words between a *syntax*-language and an *object*-language).

Church makes a distinction between an "intuitive" logic and a "formal" logic. The intuitive logic is used in speaking about the formal logic, which must be in symbolic form. Even in a given formal system, however, there seem to be two things involved. As a logic, the system may refer to things which are "logically" related. As a system of symbols, it is possible to introduce a means of speaking about the symbol itself. A symbol then has two aspects: (1) that to which it refers, (2) the symbol qua simply a symbol; or to put the matter in other words, we can use a symbol or talk about the symbol. In order to use a symbol we need but the formal rules for its manipulation; in order to talk about the symbol we must know what it is to be used for, and we must have another language (metalanguage) to be used for talking about the given language.

Chwistek, we recall, tried to develop a language which would do both and as a result his system suffers from ambiguity. Applied to logic, we have two logics: (1) the formal system whose rules and operations must be clearly stated (as Church himself admits, the formalization of this system is directed by the objects to which it is to be applied); (2) an intuitive logic which is used to talk about the formal ystem. It would appear that the attempt to formulate this intuitive ɔgic would give rise to another intuitive logic itself formalized, else ve could never talk about the logic to be formalized. Either, then, we re forced into an infinite regress, or there may be an intuitive logic ight in an absolute sense and in part, at least, unformalizable.

If this latter alternative is correct, then "it plainly implies that the vhole program of the mathematical logician is futile."[20] This, as we hall see when we later discuss the Gödel theorem, is the outcome if the rogram of the mathematical logician is the complete formalization of ɔgic. Church denies, apparently on intuitive grounds, both possi-ilities and insists that there must be a formalization or else there is no xact description, and "if there is no exact description of logic, then nere is no sound basis for supposing there is such a thing as logic."[21]

If it is true that every logical formula can be translated into an rithmetic one, then if the converse is also true, logic and arithmetic re intertranslatable and what is true of logic would also be true of rithmetic—namely, that the complete formalization of arithmetic is npossible. But, as Church points out, the set of all provable formulas ı any system of symbolic logic is enumerable. As a result the set of all ɔrmulas in any system of symbolic logic of which we can prove they are ınctions of positive integers is also enumerable, while the set of all ınctions of positive integers is not enumerable.[22] It appears that this ould indicate again that the two systems (i.e., of symbolic logic and rithmetic) are not inter-reducible even if the languages used are ıter-reducible. It appears, as a matter of fact, that each system in a ɛnse presupposes the other. Brouwer and the intuitionists had taken ıe position that neither logic nor arithmetic is reducible to the other.

---

[20] A. Church, "The Richard Paradox," *American Mathematical Monthly* Vol 41 o. 6 (1934), p. 359.

[21] *Ibid.*, p. 360.

[22] *Ibid.*, pp. 358–59.

CHAPTER ELEVEN

THE FOUNDATIONS OF MATHEMATICS

III. INTUITIONISM

The goal of intuitionist mathematics is distinct from that of formalist mathematics. The formalist endeavors to cleave a gap between the thinking of the scientist and the symbols he uses. For the formalist the symbols are self-subsistent entities—entities which can be studied apart from what they stand for. The formalist makes of his mathematics a reality in itself and unique in itself. He sees the symbol only as symbol. Not that he denies that the symbol may refer to something, but he feels that what the symbol refers to has nothing to do with the symbol itself. This is evidenced very distinctly in Church's elemental mathematics

For the intuitionist, mathematics cannot be so abstracted from the activity of thought. The intuitionist mathematician "wishes to carry on his mathematics as a natural function of the intellect—as a free, living activity of thought."[1] Further, for the intuitionist, language (whether the ordinary or formalistic), "he uses only as an aid, i.e., in order to enable himself or others to recollect his mathematical thoughts." Such a linguistic accompaniment is not a picture of mathematics and even less mathematics itself. In other words, the intuitionist accuses the formalist of confusing the symbol with what the symbol refers to.

Naturally, the distinction which is now commonplace is a valuable one to make. The language has too often been taken to refer to entities because it was apparently symbolic. And frequently statements about language are confused with statements about objects.[2] If the formalist wishes to deal with the language of mathematics he is free to do so—but he must not call his results mathematics.

As for the construction of mathematics, the most important element involved lies in the concept of unity which is the architectonic principle of the series of whole numbers.[3] (It will be recalled that Hilbert

---

[1] Cf. A. Heyting, "Die intuitionistische Grundlegung der Mathematik." *Erkenntnis*, Vol. 2 (1931), p. 106, also L. E. J. Brouwer, "Intuitionism and Formalism." *Bull. Amer. Math. Soc.*, Vol. 20 (1913), pp. 81–96.

[2] Cf. Carnap, *Philosophy and Logical Syntax;* also R. Ayer, *Language Truth and Logic*. New York, 1936.

[3] Heyting, *ibid.* The student should read, Black, *Nature of Math.*, *op. cit.*, pp 186 ff.

begins with the symbol 1 or /, and that Church's elementalistic mathematics continually uses the natural number series.) The whole numbers are to be considered merely as units which are distinguished from each other only by their position in the series. This means that for the intuitionist, the concepts of unity and order are basic to the natural number series and therefore the idea of ordinal number is prior to that of cardinal number. Numbers arising in this way, i.e., as ordered series, can have no existence apart from our thought.[4] Thus the third element in a symbolic situation—the thought for which a symbol symbolizes—is also emphasized by the intuitionists.

Mathematical objects are conditioned by human thought even though they may be independent of any particular act of thought. In so far as this is the case, a mathematical object exists only in so far as its existence can be assured by thought—only such properties can belong to the mathematical object as can be recognized by thought. This position must be kept clearly differentiated from the extreme formalist position which asserts that mathematical entities are assigned properties arbitrarily and from complete subjectivism, which asserts that these entities are ultimately subjective.

The intuitionist means that mathematical objects are inevitably bound up with the thinking act even though not created by the thinking act. My perception (in an analagous fashion) of the whiteness of this paper is clearly inevitably related to my act of perception, yet I cannot arbitrarily construct a perception of red instead of white. The natural numbers might be conceived as one of the forms (in a Kantian sense) of the mind—hence not changeable at will.[5]

Since number and its properties are so inevitably related to mind, number and its properties are affected by all the limitations of thought.[6] It follows that any supposed property of number which cannot be conceptually determined or conceptually decided cannot be a real property. In other words, the existence of a number depends upon its conceptual determinability—the existence of a property of a number depends upon the conceptual decidability, i.e., upon the actual construction of the number with its supposed property. Merely to show that "the assumption that no number with a given property exists leads to a contradiction on the basis of the given set of postulates (in this case the postulates for the number system)" would indicate the *possibility* that some number possessing that property existed—not that it *actually did so*.

[4] Cf. J. Jörgensen, *A Treatise of Formal Logic*, Vol. 3, pp. 52–53; also A. Dresden, "Brouwer's Contributions to the Foundations of Mathematics," *Bull. Amer. Math. Soc.*, Vol. 30 (1924), pp. 31 ff.

[5] Cf. Jörgensen, Vol. 3, *op. cit.*, pp. 52–53; also A. Dresden, *op. cit.*, pp. 31 ff.

[6] Cf. M. Black, *op. cit.*, p. 170.

Here lies what probably is the most distinctive difference between earlier Formalism and Intuitionism. (As we have seen, Church's conception of effective calculability really comes closer to the Intuitionists demand.) Formalism insisted that consistency implies existence whether or not you can construct the entity. Intuitionism insists that consistency implies the possibility of existence, but only when you have constructed the entity have you changed this possibility into actuality. So far then as the *existence* of new mathematical objects is concerned, they must for the intuitionists actually be constructed before we can say they exist. From the point of view of mathematics, this would really mean that infinite classes cannot be shown to exist by demonstrating that the assumption they do not exist leads to a contradiction, but only by stating a defining property. This means that an infinite class in the sense of a non-countable class becomes a meaningless concept. (Brouwer actually defines an aggregate as a *law*.[7])

Involved are the following principles:

If A, B, $C \ldots F$ are a given set of propositions, and

1. P is consistent with A, B, $C \ldots F$, then P can *possibly* be constructed from A, B, $C \ldots F$.

2. P is constructible from A, B, $C \ldots F$, then P is consistent with A, B, $C \ldots F$.

3. If P is inconsistent with A, B, $C \ldots F$, then P cannot be constructed from A, B, $C \ldots F$.

4. If P is not constructible from A, B, $C \ldots F$, then P may or may not be consistent with A, B, $C \ldots F$.

The same general argument can be applied to the concepts *true* and *false*, especially where an infinite number of entities are involved. If a proposition P is constructible from a given set of propositions A, B, $C \ldots F$, then clearly P may be said to have been *deduced* from A, $B, \ldots F$ and therefore is *true*. Every constructible proposition is therefore true. Conversely, every true proposition is constructible or is an axiom. In general, by a false proposition we mean that the contradictory of the proposition is constructible. Since every true proposition is constructible and conversely, truth and constructibility appear to be synonomous terms.

So far we have been talking about *propositions*. Consider now a propositional function $\Phi(x)$. We have two possible cases: (a) the range

[7] Brouwer, "Zur Begründung der intuitionistische Mathematik, I," *Mathematische Annalen*, Vol. 93 (1924), p. 244; also "Intuitionism and Formalism," *Bull. Amer. Math. Soc.*, Vol. 20 (1913), p. 86.

of x is finite; (b) the range is infinite. In both cases we can assert either (1) $(x)\Phi(x)$ or (2) $(Ex)\sim\Phi(x) \equiv \text{df.} \sim\Phi(x_1) \vee \sim\Phi(x_2) \vee \sim\Phi(x_3) \vee \ldots \vee \sim\Phi(x_n)$. In the case of (a) we can either construct $(x)\Phi(x)$ or else by enumeration construct that x_i for which $\sim\Phi(x_i)$. This means that we can say that either $\Phi(x_i)$ or $\sim\Phi(x_i)$, and to assert that $\sim[(x)\Phi(x)]$ means that we can construct that x_i for which $\sim\Phi(x)$ and therefore $(Ex)\sim\Phi(x)$.

But consider case (b). What do we mean when we say $\sim[(x)\Phi(x)]$ and we do not have any way of constructing the x_i for which $\sim\Phi(x)$, i.e., of discovering the $\sim\Phi(x_i)$? Obviously since in 2, there is no last $\Phi(x_n)$ we cannot enumerate the $\Phi(x_i)$. This means that we have no method of determining $\sim\Phi(x)$. Hence in this case we cannot say that $\sim[(x)\Phi(x)]$ means $(Ex)\sim\Phi(x)$, for we cannot determine (i.e., construct) that x_i for which $\sim\Phi(x)$. This means that for infinite sets we cannot say that either $(x)\Phi(x)$ or $\sim[(x)\Phi(x)]$.

Mere consistency will not do since it only implies possible constructability and therefore does not make meaningful, i.e., constructible, the proposition in question. This really means that for infinite sets the ordinary logic of two values is insufficient and we need a logic of more than two values. Included among such logics are those called modal logics.[8] The following principles would be involved in such a logic (we shall discuss modal logics later):

For infinite ranges of propositional functions: given a set of propositions or propositional functions $A, B, C \ldots F$.

1. If $f(x)$ is true, it is constructible from $A, B, \ldots F$.

2. If $f(x)$ is false, $\sim f(x)$ is constructible from $A, B, \ldots F$.

3. If $f(x)$ is consistent, it may or may not be constructible from $A, B, \ldots F$.

4. If $f(x)$ is inconsistent, then $\sim f(x)$ is either consistent or its consistency is undecidable; and if it is consistent, then it may or may not be constructible.

5. If $f(x)$ is not constructible and $\sim f(x)$ is not constructible, then $f(x)$ may be meaningless in $A, B, \ldots F$.

6. If $f(x)$ is not constructible and $\sim f(x)$ is, then $f(x)$ is inconsistent with $A, B, \ldots F$.

7. If "$f(x)$ is inconstructible" can be shown to be inconstructible,

[8] Cf. Kattsoff, "Postulational Methods I," *Phil. Sci.*, Vol. 2, p. 155. A general review of such logics is given by R. Feys, "Les Logiques Nouvelles des modalités," *Revue Néoscolastique de Philosophie*, Vol. 40 (Nov., 1937), pp. 517 ff., and Vol. 41 (May, 1938), pp. 217 ff.

then $f(x)$ is true. (This reintroduces the law of the excluded middle at a new level; otherwise this would not be a correct principle.)

In classic logic the principle which said that "every proposition was either true or false and therefore from the falsity of a false proposition one could infer the truth of that proposition" is called the *law of excluded middle.* It is this law which is at the basis of the logic of two values, truth and falsity. Thus Brouwer's argument insists upon a modification of this principle when dealing with functions whose variables have a range of an infinite number of elements.[9] Note that what is demanded is a rejection of the law of excluded middle in its usual form only where infinite sets of arguments are involved. In finite sets, the usual law is theoretically acceptable, since finite processes are possible for the determination of the functions involved.

This involves the setting up of a set of rules for what might be called Intuitionist Logic. Such a set of rules would describe the processes of a logic which re-interprets the law of excluded middle. Brouwer himself gives no such system of logic and mathematics but Heyting does. We turn to Heyting's discussion of the formal rules of such a logic.[10]

Intuitionist mathematics is an activity of thought and every language (including that of the formalists) is only an aid to communication. In principle it is impossible to set up a system of formulae equivalent to intuitionist mathematics since the possibilities of thought cannot be reduced to a finite number. The attempt to translate the most important parts of mathematics into formal language is justified by the greater precision and conclusiveness of this formal language.

The formulae of the formal system result from a finite number of axioms by the application of a finite number of rules of operation. They contain variables in addition to constants. The relation between this system and mathematics is that for a given interpretation of the constants and a given limitation on the range of the variable, every formula represents a valid mathematical proposition. If the system is so constructed that this actually is the case, then the system is consistent. This is not purely formal consistency.

---

[9] Students acquainted with the theory of functions will profit by reading Brouwer, "Uber die Bedeutung des Satzes vom ausgeschlossenen Dritten in der Mathematik, "*Journal f. d. reine und angewandte mathematik*, Vol. 154 (1925) p. 1. In this article Brouwer notes some mathematical theorems actually invalid but demonstrable on the basis of the law of excluded middle.

[10] A. Heyting, "Die formalen Regeln der intuitionistischen Logik," *Sitzungsberichte d. Preussischen Akademie der Wissenschaften, Phys-Math. Klasse*, 1930, pp. 42 ff., pp. 57 ff., and pp. 158 ff.; also Heyting, "Die intuitionistische Grundlegung," *op. cit.*, pp. 113 ff., and Heyting, "Math. Grundlagenforschungen," *Ergebnisse der Mathematik und ihrer Grenzgebiete*, Vol. 3, No. 4, pp. 13 ff.

1. RULES OF OPERATION[11]

1.1 The symbol $\vdash$ denotes that a formula is included in the list of "valid formulae." If the formula is an axiom, i.e., *taken* to be valid, then we denote this by $\Vdash$.

1.2 If a and b are valid formulae, $a \wedge b$ is also ($a \wedge b$ is read "a and b").

1.3 If a and $a \supset b$ are valid formulae, then b is also. ($a \supset b$ is read "from a follows b").

1.4 The expression "Const a" denotes that a is a constant. Every symbol not introduced by the expression "Const" is a variable. From a valid formula we always get another valid formula by replacing a given variable wherever it occurs by a given combination of symbols.

1.5 $\binom{p}{x}a$ is the formula which results from formula a whenever we replace variable x (if it occurs in a) everywhere by the symbol-combination p. (Note that the direction of substitution is the opposite of that used by Church.)

1.6 The formula $a =_{Df.} b$ is called a *definition*. It denotes that a valid formula results from another valid formula if we replace a by b (or conversely b by a) even if this is not done everywhere.

Actually the intuitionist must distinguish between a sentence (Aussage) and a proposition(Satz).[12] A proposition is the assertion (Behauptung) of a sentence. A mathematical sentence expresses a definite expectation. For example, the sentence "Euler's constant C is a rational number" expresses the expectation that we can find two whole numbers a and b such that $C = \frac{a}{b}$. The assertion of a sentence (i.e., a proposition) denotes that the expectation has been fulfilled, i.e., the assertion that C is a rational number means that a and b have actually been found. $\vdash$ denotes the proposition, not the sentence.

It is important to note that in the formalist system, the proposition may be viewed as a sentence which can be true or false. Also, according to the formalist the assertion that C is a rational number may not involve the actual construction of a and b. The intuitionist demands their actual discovery. Church's notion of effective calculability is in this respect close to the demand of the intuitionist.

---

[11] Heyting, "Die formalen Regeln," *op. cit.*, p. 45.
[12] Heyting, "Die intuitionistische Grundlegung," *op. cit.*, p. 113.

2. CONST $\wedge$

(may be read "and") ., :, :·, ::, $\supset$, $\wedge$, $\supset\subset$, (,), $=_{Df}$.

2.1 $\vdash\!\vdash . a \supset a \wedge a$

2.11 $\vdash\!\vdash . a \wedge b \supset b \wedge a$

2.12 $\vdash\!\vdash . a \supset b . \supset . a \wedge c \supset b \wedge c$

2.13 $\vdash\!\vdash . a \supset b . \wedge . b \supset c : \supset : a \supset c$

2.14 $\vdash\!\vdash . b \supset . a \supset b$

2.15 $\vdash\!\vdash . a \wedge . a \supset b . \supset b$

2.01 $\vdash . a \supset\subset b . =_{Df} . a \supset b . \wedge . b \supset a$

2.02 $\vdash . a \wedge b \wedge c =_{Df} . a \wedge b . \wedge c$

We shall give only theorems without proof here, since this does not differ from the formalist system so far.

2.2 $\vdash . a \wedge b \supset a$

2.21 $\vdash . a \supset a$

2.22 $\vdash . a \wedge b \supset b$

2.23 $\vdash . a \supset b . \wedge . c \supset d : \supset : a \wedge c \supset b \wedge d$

2.24 $\vdash . a \supset b . \wedge . a \supset c : \supset\subset : a \supset b \wedge c$

2.25 $\vdash . b . \wedge . a \supset c : \supset : a \supset b \wedge c$

2.26 $\vdash . b \supset . a \supset a \wedge b$

2.27 $\vdash : a \supset . b \supset c : \supset\subset : a \wedge b \supset c$

2.271 $\vdash : a \supset . b \supset c : \supset\subset : b \supset . a \supset c$

2.28 $\vdash . a \supset c . \supset . a \wedge b \supset c$

2.281 $\vdash : a \supset b . \supset : a \supset . c \supset b$

2.282 $\vdash . a \wedge . a \wedge b \supset c . \supset . b \supset c$

2.29 $\vdash : a \supset b . \supset : b \supset c . \supset . a \supset c$

2.291 $\vdash : b \supset c . \supset : a \supset b \supset . a \supset c$

2.3 $\vdash . a \wedge b . \wedge c . \supset . a \wedge . b \wedge c$

2.31 $\vdash . a \wedge b . \wedge c . \supset . b \wedge a . \wedge c$

2.32 $\vdash . a \wedge . b \wedge c . \supset . a \wedge b . \wedge c$

2.33 $\vdash . a \wedge b \wedge c \supset\subset a \wedge c \wedge b \supset\subset b \wedge a \wedge c$ etc.

2.4 $\vdash . a \supset b . \wedge . b \supset c . \wedge . c \supset d . \supset . a \supset d$

3. CONST $\vee$ (May be read "either-or")

3.1 $\vdash\vdash . a \supset a \vee b$

3.11 $\vdash\vdash . a \vee b \supset b \vee a$

3.12 $\vdash\vdash . a \supset c . \wedge . b \supset c : \supset : a \vee b \supset c$

3.01 $\vdash . a \vee b \vee c =_{Df.} . a \vee b . \vee c$

3.2 $\vdash . a \vee b . \vee c . \supset . a \vee . b \vee c$

3.21 $\vdash . a \vee b \vee c . \supset . a \vee b . \vee c$

3.22 $\vdash . a \vee a \supset a$

3.3 $\vdash . a \supset b . \wedge . c \supset d . \supset . a \vee c \supset b \vee d$

3.31 $\vdash . a \supset b . \supset . a \wedge c \supset b \vee d$

3.32 $\vdash . a \supset b . \supset . a \vee b \supset b$

3.33 $\vdash . a \vee b \supset b . \supset . a \supset b$

3.34 $\vdash . a \supset b . \supset . a \vee c \supset b \vee c$

3.35 $\vdash . a \vee b \vee c \supset b \vee a \vee c \supset a \vee c \vee b$, etc.

3.4 $\vdash . a \wedge c . \vee . b \wedge c . \supset . a \vee b . \wedge c$

3.41 $\vdash . a \vee b . \wedge c . \supset . a \wedge c . \vee . b \wedge c$

3.42 $\vdash . a \wedge b . \vee c . \supset\subset . a \vee c . \wedge . b \vee c$

3.5 $\vdash : a \supset b \vee c . \wedge . b \supset d . \wedge c \supset e : \supset : a \supset d \vee e$

3.51 $\vdash : a \supset b \vee c . \wedge . b \supset d . \wedge . c \supset d : \supset : a \supset d$

3.6 $\vdash : a \vee b \supset : a \supset b . \supset b$

This is one place where this system departs from the formalist logic. The converse of this is not valid. Its proof depends on 3.32; 2.271.

4. CONST $\urcorner$ (May be read "not." This is not to be identified with $\sim$.)

This is *negation.* Its properties differ from the $\sim$ of formalist logic. The four concepts now introduced are $a \supset b$, $a \wedge b$, $a \vee b$, $\urcorner a$. Contrary to formalist logic, these cannot be defined in terms of each other. They are irreducible.

A loose interpretation of the $\neg a$ may be "a is absurd." Thus $\neg\neg a$ will be, "it is absurd, that a is absurd." A little thought will then show that $\neg\neg a$ is not the same as saying that "a is true" as it would be in the case of "it is false that a is false," or $\sim(\sim a)$ of the two-valued logic. Or again, if we consider $\neg a$ to read "the truth of a cannot be determined in a finite number of steps," then $\neg\neg a$ which would read "the truth of the nondeterminability of the truth of a in a finite number of steps cannot be determined in a finite number of steps," and this would not mean that the truth of a could be determined in a finite number of steps. So while for the formalist logic $\sim\sim a = a$, for the intuitionist logic $\neg\neg a \neq a$. It will be noticed that in all proofs involving only finite elements, the two logics are the same.

4.1 $\vdash\!\vdash . \neg a \supset . a \supset b$

4.11 $\vdash\!\vdash . a \supset b . \wedge . a \supset \neg b . \supset \neg a$

4.01 $\vdash . \neg\neg a =_{Df.} \neg(\neg a)$

4.1 says "if not-a, then a implies any b." 4.11 says "if a implies both b and not-b, then not-a."

4.2 $\vdash . a \supset b . \supset . \neg b \supset \neg a$

The proof follows from 2.14, 2.12, 4.11, and 2.27. This is the law of contradiction and interchange.

4.21 $\vdash . a \supset \neg b . \supset . b \supset \neg a$

4.22 $\vdash . a \supset b . \supset . \neg\neg a \supset \neg\neg b$

4.23 $\vdash . a \wedge b \supset c . \supset . a \wedge \neg c \supset \neg b$

4.24 $\vdash . a \supset b . \wedge \neg b . \supset . \neg a$

4.3 $\vdash . a \supset \neg\neg a$

Proof follows from 2.21 and 4.21. The converse is not demonstrable here.

4.31 $\vdash . \neg a \supset \neg\neg\neg a$

4.32 $\vdash . \neg\neg\neg a \supset \neg a$

4.4 $\vdash . a \wedge \neg a \supset b$

4.41 $\vdash . a \wedge \neg a . \vee b \supset b$

4.42 $\vdash . a \vee b . \wedge \neg a . \supset b$

4.43 $\vdash . \neg a \supset . \neg b \supset \neg(a \vee b) .$

4.44 $\vdash . \neg(a \vee b) \supset\subset \neg a \wedge \neg b$

4.45 $\vdash . a \vee \neg a . \supset . \neg\neg a \supset a$

4.46 $\vdash . \neg a \vee b . \supset . a \supset b$

4.47 $\vdash . a \vee b . \supset . \neg a \supset b$

The converses of 4.46 and 4.47 are also not demonstrable.

4.5 $\vdash . b \wedge a \supset \neg a . \supset . b \supset \neg a$

4.51 $\vdash . a \wedge \neg(a \wedge b) \supset \neg b$

4.52 $\vdash . \neg(a \wedge b) \supset . a \supset \neg b$

4.521 $\vdash . a \supset \neg b . \supset \neg(a \wedge b)$

4.53 $\vdash . \neg a \vee \neg b \supset \neg(a \wedge b)$

4.54 $\vdash . \neg(a \wedge b) \wedge . a \vee \neg a . \supset \neg a \vee \neg b$

4.6 $\vdash . \neg\neg a \wedge \neg\neg b \supset \neg\neg(a \wedge b)$

4.61 $\vdash . \neg\neg(a \wedge b) \supset \neg\neg a \wedge \neg\neg b$

4.62 $\vdash . \neg\neg a \vee \neg\neg b \supset \neg\neg(a \vee b)$

4.63 $\vdash . \neg\neg(a \vee b) \wedge . \neg a \vee \neg\neg a . \supset \neg\neg a \vee \neg\neg b$

4.7 $\vdash . a \supset \neg(b \wedge c) . \supset . a \wedge b \supset \neg c$

4.8 $\vdash . \neg\neg(a \vee \neg a)$

This is the intuitionist form of the law of excluded middle. The formalist form is $a \vee \sim a$. This is an important difference in approach. In this 4.8 lies the reinterpretation of the law of excluded middle. The proof of 4.8 is based on . 3.1, 4.2, 3.11, and 4.11.

4.81 $\vdash . \neg\neg(\neg\neg a \vee a)$

4.82 $\vdash . a \vee \neg a \supset b . \supset . \neg\neg b$

4.9 $\vdash . a \supset b . \supset \neg(a \wedge \neg b)$

4.91 $\vdash . a \vee b \supset \neg(\neg a \wedge \neg b)$

4.92 $\vdash . a \wedge b \supset \neg(\neg a \vee \neg b)$

The converses of 4.9, 4.91 and 4.92 cannot be demonstrated.

This gives in outline the foundation of intuitionist logic. On this basis we now set up the formal rules of intuitionist mathematics.[13]

---

[13] Cf. Black, *Nature of Math.*, pp. 201 ff. Cf. also S. C. Kleene, "On the Interpretation of Intuitionistic Number Theory," *Journal of Symbolic Logic*, Vol. 10, (1945), pp. 109–24.

The actual construction of mathematics begins with the consideration of objects (natural numbers, aggregates). But for the formal manipulation it is more convenient to start with general propositions concerning relations between objects, and formulae containing variables for entities.

The fundamental relation between individuals is $p \equiv q$; read "p is the same object as q." $p \equiv p$ is true of objects only, and is, indeed, equivalent to the statement, "p is an object."

$=$ is used for cardinal equality. This can relate objects not identical.

$\equiv$ is used for mathematical identity.

The fundamental ideas of the functional calculus as developed in an earlier chapter can be transferred, except that (x) cannot be defined in terms of Ex and Ex cannot be defined in terms of (x).

There is also a fundamental difference in the interpretation of the universal quantifier and the existential quantifier. The expression "there exists a number n having the property P" is to be taken to mean "a number n can actually be given which has the property P."

5. RULES OF OPERATION

5.1 Variables are divided into propositional variables and object-variables. Propositional variables are denoted by $p, q, r \ldots$, object-variables by $a, b, c \ldots$.

5.2 The relations $\binom{p}{x}$, (x), (Ex) where x is an object-variable, make x in that part of the formula to which they refer, into a *bound variable.* Such a variable must be replaced only by another object-variable.

Definition of "expression" (Ausdruck):

5.3 Every valid formula is an expression.

5.31 If we replace an object in an expression everywhere by another object we get another expression.

5.32 If in an expression, we replace a propositional variable everywhere it occurs by an expression or an object variable is replaced everywhere it occurs by a symbol combination p for which $p \equiv p$ is valid, then we get another expression.

5.33 If we replace a variable in an expression everywhere it occurs by another variable of the same kind, then we get another expression.

5.34 If A is an expression $\neg A$ is also and conversely.

5.35 If A is an expression in which x does not occur as a bound variable, then $\binom{p}{x}A$, $(x)A$, $(Ex)A$ are expressions.

5.36 If A and B are expressions, then $A \supset B$, $A \wedge B$, $A \vee B$ are expressions.

5.37 In an expression, the range of two bound symbolic combinations such as are named in 5.2, which contain the same variables, must not overlap.

5.4 A valid formula results from a valid formula if we replace a propositional variable by an arbitrary expression or an object-variable by a symbolic combination p, for which $p \equiv p$ is a valid formula.

5.5 If $a \supset b$ and $(p) . \binom{p}{x} a \supset\subset a$ are valid formulae then $a \supset (x)b$ is a valid formula.

5.6 If $a \supset b$ and $(p) . \binom{p}{x} b \supset\subset b$ are valid formulae then $(Ex)a \supset b$ is valid.

5.7 If in the expression A the variable x occurs only as a bound variable then $(y) . \binom{y}{x} A \supset\subset A$ is a valid formula.

5.8 If A is a valid formula which does not contain x as a bound variable, then $(x)A$ is a valid formula.

Proof: $\vdash A \supset [2.14] \vdash 1 \equiv 1 \supset A . \supset . [5.5] \vdash 1 \equiv 1 \supset (x)A . \supset .$ $[1.3] \vdash (x)A.$

6. CONST 1, $\equiv$, (E), $.(.)$, (Fundamental concepts), (—), (abbreviation)

6.1 $\vdash\vdash . 1 \equiv 1$

6.11 $\vdash\vdash . p \equiv . p \wedge q . \equiv q : \supset : p \equiv q \vee \neg(p \equiv q)$

6.12 $\vdash\vdash . q \equiv p . \supset . p \equiv q$

Axioms concerning $\binom{p}{x}$

6.2 $\vdash\vdash . \binom{x}{x} a \supset\subset a$

6.21 $\vdash\vdash . \left(\binom{q}{x} p \atop x \right) a \supset\subset \binom{q}{x} \binom{p}{x} a$

6.22 $\vdash\vdash . \binom{p}{x} . a \supset b : \supset : \binom{p}{x} a \supset \binom{p}{x} b$

6.221 $\vdash\vdash . \binom{p}{x} a \supset \binom{p}{x} b : \supset : \binom{p}{x} . a \supset b$

6.23 $\vdash\vdash . \binom{p}{x} a \wedge \binom{p}{x} b : \supset : \binom{p}{x} . a \wedge b$

6.24 $\vdash\vdash . \binom{p}{x} . a \vee b : \supset . \binom{p}{x} a \vee \binom{p}{x} b$

6.25 $\vdash\vdash . \neg \binom{p}{x} a \supset \binom{p}{x} \neg a$

6.26 $\vdash\vdash . p \equiv q \supset . \binom{p}{x} a \supset \binom{q}{x} a$

Axioms concerning (x)

6.3 $\vdash\vdash . (x) a \supset \binom{p}{x} a$

6.31 $\vdash\vdash . (x) a \supset (y) \binom{y}{x} a$

6.32 $\vdash\vdash . (y) . y \equiv y \supset \binom{y}{x} a : \supset : (x) a$

6.01 $\vdash . g(\bar{x}) =_{Df.} . (p) \binom{p}{x} g \equiv g . \vee . (p) \binom{p}{x} g \supset\subset g$

The first alternative holds when g is an object, the second when it is an expression. $g(\bar{x})$ denotes: "g does not contain x." The x is therefore a bound variable.

We note some theorems without proof.

6.4 $\vdash . (x) . a \supset b : \supset : (x) a \supset (x) b$

6.411 $\vdash . (x) a \wedge (x) b \supset : (x) . a \wedge b$

6.413 $\vdash . a(\bar{x}) \wedge b(\bar{x}) \supset . a \supset b . (\bar{x})$

6.43 $\vdash . (x) \neg a \supset \neg (x) a$

6.431 $\vdash . \binom{p}{x} \neg a \supset \neg \binom{p}{x} a$

6.44 $\vdash . p(\bar{x}) \supset . \binom{p}{x} a . (\bar{x})$

Axioms concerning (Ex)

6.5 $\vdash\vdash . \binom{p}{x} a \supset (Ex) a$

6.51 $\vdash\vdash . a(\bar{x}) \supset . (Ex)a \supset (x)a$

6.52 $\vdash\vdash . (x) . a \supset b . \supset . (Ex)a \supset (Ex)b$

Some theorems.

6.6 $\vdash . (x)a \supset (Ex)a$

6.61 $\vdash . a(\bar{x}) \supset . \binom{p}{x} a \supset (x)a$

6.7 $\vdash . (x)\neg a \supset \neg(Ex)a$

6.71 $\vdash . \neg(Ex)a \supset (x)\neg a$

6.72 $\vdash . (Ex)a \supset \neg(x)\neg a$

6.73 $\vdash . (x)a \supset \neg(Ex)\neg a$

6.74 $\vdash . (Ex)\neg a \supset \neg(x)a$

6.75 $\vdash . \neg(x)\neg a \supset \neg\neg(Ex)a$

6.76 $\vdash . \neg(Ex)\neg a \supset (x)\neg\neg a$

6.77 $\vdash . (Ex)\neg\neg a \supset \neg\neg(Ex)a$

6.78 $\vdash . \neg\neg(x)a \supset (x)\neg\neg a$

7. CONST ϵ, ‘‘, (fundamental concepts)
$\textcircled{1}$,’’ 0 (abbreviations)

ϵ is read "is a . . ." in the sense of "is an element in the class" or "has the property." It must not be confused with "is a subclass of." $x‘p$ is read "the x of p."

$q\textcircled{1}p$ denotes a *unique image* of p on q. (For the sake of printing, I have changed Heyting's symbols to $\textcircled{1}$, 0.) p is its range.

x’’r denotes the *species*[14] of the images of the elements of a subspecies r of the range of x.

$x0r$ is the correlation which has as its range r but correlates the same images to the elements of r as x (mapping all of p and q) does.

7.01 $\vdash . x\epsilon q\textcircled{1}p =_{Df.} . (r) . r\epsilon p \supset \subset x‘r\epsilon q$

7.02 $\vdash : x\epsilon q\textcircled{1}p . \wedge . (y) . y\epsilon r \supset y\epsilon p . \supset : s\epsilon x’’r =_{Df.} .$
$(Et) . t\epsilon r \wedge x‘t \equiv s.$

7.03 $\vdash : x\epsilon q\textcircled{1}p \wedge . (y) y\epsilon r \supset y\epsilon p \supset : z \equiv x0r =_{Df.} . z\epsilon q\textcircled{1}r \wedge . (s) . s\epsilon r \supset$
$z‘s \equiv x‘s.$

---

[14] The term species (spezies) is defined as a property. It is like the term class of formalist development. Cf. Brouwer, *Math. Ann.* 93, p. 245, and below.

7.04 $\vdash . x`y`p =_{Df.} . x` . y`p$

7.041 $\vdash : x, y \epsilon q \textcircled{1} p : \supset : x \equiv y . =_{Df.} . (r) . r \epsilon p \supset x`r \equiv y`r \vee x`r \equiv y`r$

7.1 $\vdash\vdash : x \epsilon s \textcircled{1} r \wedge p, q \epsilon r \wedge p \equiv q \supset . x`p \equiv x`q$

7.11 $\vdash\vdash . x \epsilon q \textcircled{1} p \supset x \equiv x$

7.12 $\vdash\vdash : x \epsilon q \textcircled{1} p \wedge x \equiv y \wedge r \epsilon p \supset . x`r \equiv y`r$

7.05 $\vdash . x, y \epsilon p =_{Df.} x \epsilon p \wedge y \epsilon p$

7.06 $\vdash . x, y, z \epsilon p =_{Df.} x \epsilon p \wedge y \epsilon p \wedge z \epsilon p$

10. CONST $\mathcal{N}$, seq (fundamental concepts)
$\equiv$, $=$, $+$, $-$, $>$, $<$, (abbreviations)

$\mathcal{N}$ denotes "natural number."

Seq‘p denotes "the number following p."

10.1 $\vdash\vdash . 1 \epsilon \mathcal{N}$

10.11 $\vdash\vdash . 1 \equiv 1$

10.12 $\vdash\vdash . \text{seq} \epsilon \mathcal{N} \textcircled{1} \mathcal{N}$

10.13 $\vdash\vdash . p \epsilon \mathcal{N} \supset \neg(\text{seq}`p \equiv 1)$

10.14 $\vdash\vdash . \binom{1}{x} a \wedge . (p) . p \epsilon \mathcal{N} \wedge \binom{p}{x} a \supset \binom{\text{seq}`p}{x} a . \supset . (q) . q \epsilon \mathcal{N} \supset \binom{q}{x} a$

10.15 $\vdash\vdash . p, q \epsilon \mathcal{N} \wedge \text{seq}`p \equiv \text{seq}`p \supset . p \equiv q$

10.14 is the fundamental proposition of complete induction.

10.01 $\vdash . p, q \epsilon \mathcal{N} \supset . p = q =_{Df.} p \equiv q =_{Df.} . p \equiv q$

This states that for natural numbers p and q, the three symbols, $=$, $\equiv$, $\equiv$, are by definition the same. For example, $p = q$ is by definition $p \equiv q$, which is by definition $p \equiv q$.

10.03 $\vdash . p \epsilon \mathcal{N} \supset . p + 1 =_{Df.} \text{seq}`p$

10.04 $\vdash . p, q \epsilon \mathcal{N} \supset . p + \text{seq}`p =_{Df.} \text{seq}`(p + q)$

10.05 $\vdash . p, q \epsilon \mathcal{N} \supset . r = p - q =_{Df.} q + r = p$

10.06 $\vdash . p, q \epsilon \mathcal{N} \supset . p > q =_{Df.} q < p =_{Df.} p - q \epsilon \mathcal{N}$

This means that, for p and q natural numbers, "p is greater than q" means by definition "q is less than p," which is by definition "p-q is also a natural number."

Some theorems:

10.24 $\vdash . p \epsilon \mathcal{N} \supset . p = q . \supset\subset : \mathrm{seq}`p = \mathrm{seq}`q$

10.4 $\vdash . p,q \epsilon \mathcal{N} \supset p + q \epsilon \mathcal{N}$

10.41 $\vdash . p,q,r \epsilon \mathcal{N} \wedge p = q \supset . p + r = q + r$

10.42 $\vdash . p,q,r \epsilon \mathcal{N} \supset . (p+q) + r = p + (q+r)$

10.43 $\vdash . p \epsilon \mathcal{N} \supset . p + 1 = 1 + p$

10.431 $\vdash . p,q \epsilon \mathcal{N} \supset . p + q \equiv q + p$

10.5 $\vdash . p \epsilon \mathcal{N} \supset \mathrm{seq}`p > p$

10.55 $\vdash . p \epsilon \mathcal{N} \wedge \binom{1}{x} a \wedge . (q) . q \epsilon \mathcal{N} \wedge q < p \wedge \binom{q}{x} a \supset \binom{\mathrm{seq}`q}{x} a : \supset :$

$$(r) . r \epsilon \mathcal{N} \wedge r < p + 1 \supset \binom{r}{x} a$$

The symbolic development of Heyting becomes much more complicated as it approaches the definition of aggregates, species, etc., so we shall follow Brouwer from this point.[15]

At the foundation of mathematics, says Brouwer, there lies an unlimited series of symbols which are defined by a first element and the law which derives from each element of this series of symbols the next one in the series. The sequence of numbers ("Nummern") ζ i.e., 1, 2, 3, 4, 5, . . . is an especially useful series of this kind.

An *aggregate* (Menge) is a *law* such that if arbitrary numbers are chosen in succession, each choice either generates a definite series of symbols and does or does not end the process, or causes the process to be stopped and the result to be definitely rejected. In either case for every $n > 1$ it is possible to give after every unending and unchecked sequence of $n-1$ choices at least one number which if selected as the nth number does *not* cause the process to be checked. Every sequence of series of symbols generated by an unending choice-sequence in this way is called an *element of the aggregate*.

This rather complicated definition of aggregate is not a definition in extension. An aggregate is *not* the totality of its elements but the law which defines it. Actually, as Heyting points out, two laws are involved—the first limiting the freedom of choice and the second correlating definite mathematical entities to the finite sequence of permissible choices.[16] The first law must be such that (1) after n choices

[15] We now follow Brouwer, "Zur Begründung der intuitionistischen Mathematik I," *Mathematischen Annalen*, Vol. 93 (1925), pp. 244 ff., and II, *ibid.*, Vol. 95 (1925), pp. 453 ff.

[16] Heyting, *Math. Grundl.* p. 24; cf. also Erkenntnis, *op. cit.*, Vol. 2, p. 110; and K. Menger, "Bemerkungen zu Grundlagenfragen," *Jahresb. d. Deutschen Math.-Verein.*, Vol. 37 (1928), pp. 219 ff.

it is determined for every natural number whether or not it can be the $(n+1)$st choice; (2) for every $n>1$ after every sequence of $n-1$ permissible choices a number can be given which can be the nth choice. The second law correlates a sequence of permissible choices with a sequence of mathematical objects which are called elements of the aggregate.

An aggregate is said to be *finite* (finit) if for every n in ζ, a number k_n is determined such that whenever a number lying in ζ higher than k_n, is chosen for the nth choice the process is checked.

If different unchecked choice-sequences always lead to different sequences of series of symbols, the aggregate is said to be *individualized.*

We shall denote the aggregate of numbers (i.e., of the symbol-series of ζ) by A.

Two elements of an aggregate are said to be *equal* or *identical* if we are sure that for every n the nth choice for the two elements generates the same series of symbols.

The aggregate M is said to be a *subaggregate* of N if for every element of M there exists an element equal to it in N.

Aggregates and their elements are called *mathematical entities.*

By a *species of the first order* we mean a property which only a mathematical entity can have. Such an entity is called an *element of the first order species.*

The property "divisible by 2," which defines the aggregate of even numbers, is a species of the first order. The entity is then an element of the species of the first order. It is important to note that Brouwer's development is in terms of properties and laws. The aggregate was defined as a law. The species of various orders are also defined in terms of properties. In other words, Brouwer is viewing these concepts not so much in terms of the elements subsumed under them (extension) as in terms of laws selecting the elements (intension). The assertion of "existence" of an element will then mean the demonstration that the law actually selects such an element. In the case of an infinite class of elements, the failure to select the element makes impossible the assertion of the existence of the element.

By a *species of the second order* we mean a property which can be predicated only of a mathematical entity or species of the first order.

In an analagous fashion we can define a species of the nth order. A species of the nth order is a property applicable only to species of $(n-1)$st order. This defines properties applicable to entities, properties applicable to properties of entities, etc.

Species M is said to be a subspecies of species N if for every element of M there exists an element of N equal to it.

Two aggregate elements (species) are said to be *different* if the impossibility of their equality is definite.

A species such that of any two of its elements it can be determined whether they are equal or different is called a *discrete* species.

Species M is said to *project beyond* species N if M contains an element different from every element in N.

Two species are said to be *congruent* if neither projects beyond the other, i.e., when every property impossible for the elements of one species is impossible for the elements of the other.

The species which contains those elements which belong to M *and* to N is called the *product* (Durchschnitt) of M and N and is written $D(M,N)$.

The species which contains those elements which belong *either* to species M *or* to species N is called the *sum* (Vereinigung) of M and N and is written $S(M,N)$.

A subspecies of N congruent with M is also said to be *half-identical* with N.

Two species are said to be *nonoverlapping* if they are different and it is impossible for an element of M to exist which is identical with an element of N.

If M' and M'' are nonoverlapping subspecies of N and if $S(M',M'')$ is half-identical with N, then N is said to be *composed* of M' and M''. M' and M'' are called *complementary* species in N.

If M' and M'' are nonoverlapping subspecies of N and $S(M',M'')$ is identical with N, then we say N is *divided* into M' and M''. M' and M'' are called *conjugate division* species of N. M' as well as M'' is called also a *separable* subspecies of N.

Whenever we can set up a one-to-one relation, i.e., a law between two species M and N which correlates every element of M with an element of N in such fashion that equal and only equal elements of M correspond to equal elements of N and every element of M is correlated with an element of N, then we write $M \sim N$ and say that M and N have the same *potency* or *cardinal number*.

A species is said to be *finite* (endlich) if it has the same potency as an initial segment of the sequence ζ.

A species is said to be *infinite* if it contains a subspecies of equal potency with A (defined above). It is said to be *reducibly* infinite if the subspecies in question is separable.

There is no reason to assert that every species or aggregate is either finite or infinite but no species can be finite and infinite at the same time.

Two species M and N (whose cardinal numbers are m and n respectively) are said to be *equivalent* if every element of M is correlated by a law G_1 to such an element of N that equal and only equal elements of M correspond to equal elements of N, and every element of N is correlated by a law G_2 to such an element of M that equal and only equal elements of N correspond to equal elements of M. We denote this property by $m=n$.

If to each element of M such an element of N is correlated that to equal and only to equal elements of M there correspond equal elements of N, but no law exists for the converse correlation, then we write $m<n$ or $n>m$ and say that N is *greater* than M and M is *less* than N.

This is sufficient for our purposes. For further development within mathematics itself the reader is referred to the works of Brouwer cited earlier. An excellent bibliography of the more technical mathematical investigations from the intuitionist point of view is given in Heyting's *Mathematische Grundlagenforschungen.*

The amazing thing about the Intuitionist–Formalist controversy is that it is really a problem not solvable mathematically, but logically and epistemologically: The fundamental difference between the two schools ultimately reduces to the interpretation of the law of excluded middle. This gives the variations of the nature of existence—namely, for the intuitionists a mathematical entity exists only if it can be constructed; for the formalists it exists (a) if it can be constructed or (b) if its nonexistence is inconsistent.

The reinterpretation of the law of excluded middle leads to logics of more than two values (the so-called "modal logics"[17]) and hence there is involved in Intuitionism a multi-valued logic. The formalists have come closer to the intuitionists on the question of constructability, but not on the use of a multi-valued logic. Both Hilbert and the intuitionists really agree in asserting the impossibility of deducing mathematics from formal logic.[18]

The difference between Intuitionism and Logistics lies in the same logical problem and also in the question whether the number sequence is given in intuition or derived from logical concepts.

In addition, Intuitionism differs from both schools in asserting the connotative significance of mathematics over and above its purely formal significance. The intuitionists wish to provide for the explanation of the application of mathematics. As Heyting points out, there are divergences among the intuitionists but they all agree in two fundamental propositions:

1) Mathematics has not only a formal significance but a connotative one as well.

2) Mathematical objects are grasped by the thinking spirit and mathematics is therefore independent of experience.[19]

Brouwer gives four points which will bridge the gap between him-

[17] See section on Becker's article "Zur Logik der Modalitaten," listed below.

[18] Cf. for a discussion of the identity and difference between the two schools, O. Becker, "Mathematische Existenz," *Jahrbuch für Philosophic und phänomologische Forschung*, Vol. 8, 1927.

[19] Heyting, "Mathematische Grundlagenforschungen," p. 3.

self and Hilbert whenever they are solved or at least adequately expounded.[20]

1. The distinction between the formalistic efforts to construct the whole of mathematics as a body of formulae and an intuitive (connotative) theory of the laws of this structure as well as the recognition that for formalism the intuitionist mathematics of the aggregate of natural numbers is indispensable.

2. The rejection of the uncritical use of the logical law of excluded middle as well as the knowledge that the investigation of the reason for the customary acceptance of that law and its sphere of applicability is an important part of the investigations of the foundations of mathematics.

3. The identification of the law of excluded middle with the principle of the solvability of every mathematical problem.

4. The recognition that the (connotative) justification of formalist mathematics by means of a consistency proof contains a vicious circle.

Although for a time the attack on Brouwer seemed to be successful (it was argued that Brouwer's position entailed catastrophic consequences for mathematics), recent investigations by the formalists have demonstrated that his suspicion of mere consistency was justified. It has been shown (by Gödel) that mathematics can never be completely formalized; that in every arithmetic system it will be possible to construct propositions which cannot be decided in that system, i.e., propositions such that it is impossible to decide whether they are true or false. This does not mean a temporary lack of proof but a permanent one so far as the given system is concerned.[21] This itself it seems to me leads to the necessity of a three-valued logic. Every proposition will be either true, or false, or undecidable. In such a logic we would say: if a proposition is not undecidable, it is either true or false.

If a proposition is not true, it is either false or undecidable.

If a proposition is not false, it is either true or undecidable.

The use of proofs involving the law of excluded middle, is possible then only with care. Brouwer is shown right by formalist investigations.

It must be remembered, however, that when we are in a realm where finite constructions are possible, the ordinary two-valued logic holds good. The two schools are in agreement on this point. The

[20] Brouwer, "Intuitionistische Betrachtungen über den Formalismus," *Sitzungs b. d. Preuss. Akad. d. Wiss. Phys.-Math. Klasse*, 1928, pp. 49 ff.

[21] Cf. F. Waismann, *Einfuhrung in das Math. Denken*, pp. 80–91.

schools are at variance only in the realm of the infinite. For this reason the intuitionists have been called *Finitists.* Let us consider the position of Finitism in more detail.[22]

An illustration of the type of statement where the two schools differ is the Fermat theorem: "for all $n>2$, $x^n+y^n \neq z^n$, is true" or else the proposition, "there exist an x, y, and z such that for $n>2$, $x^n+y^n=z^n$" is true. Since these two propositions are contradictory the logistic position is that the law of excluded middle *always* holds.

Furthermore, the logisticians would hold, if we could demonstrate the inconsistency of the universal form, we could infer the existential form. Brouwer and the finitists, however, would insist that where you have an infinite range of objects the two propositions stated are not real contradictories, and hence from the inconsistency of one you cannot infer the other. (The reader should recall that the paradox of the class of all classes not containing themselves as members is a situation of the sort described. This limitation by the finitists would, therefore, avoid the paradoxes of this type with the necessity of a theory of types, etc.)

Actually the finitist is saying that if conditions are such that the truth or falsity of a proposition cannot be decided at all, then it is nonsense to say that the proposition must still be true or false.[23] Such conditions exist where the range of the subject class is infinite, so that its members cannot be exhibited individually and the general proposition concerning the members is not known to be true or false. Obviously there is involved here a variation in the meaning of *true* and *false* as well as in the law of excluded middle.

Let us consider another example. "All numbers of the form $2^{2n+9}+1$ are factorable or else there exists a number of this form which is prime."[24] Suppose we cannot show that the property *factorable* belongs to all numbers of this form on the ground of the axioms. Then since there is no way of producing all numbers of this form, the general statement can never be verified. As for the existential fact here one might by *chance* verify it, but one could never prove it false. The finitist therefore asserts that as a result the alternants posed are neither true nor false. Such forms are not completely meaningful.

---

[22] A. Ambrose, "Finitism in Mathematics," *Mind*, Vol. 44 (1935), pp. 186–203 and pp. 317–40.

[23] Two methodological positions arise from such a principle: (1) Logical Positivism, insisting that questions whose answers are not capable of empirical verification are meaningless; (2) Operationalism, insisting that the operations involved in a proposition determine the meaning of the proposition. For logical positivism see Ayer, "Language, Truth, Logic" and Weinberg, "An Examination of Logical Positivism."

[24] Ambrose, pp. 190–91. Cf. also A. Reymond, "La négation et le principe du tiers exclu.," *Actes du congres intern. de Phil. Scientifique*, VI, Phil. des Math. Paris, 1936, pp. 62 ff.

The finitist position influences also the consideration of integrals which can be evaluated only by approximation methods. Such integrals do not define numbers for which it can be decided whether they are rational or irrational—since obviously the question cannot be decided unless by chance we discover it equal to some rational number.

The difficulty lies of course in the form of "not-p," so far as the case p (here p is a proposition of the form "all . . .") is concerned. The formal negative of "all" is "not all." Then the question is "Do the symbols for *not all* and *there exist* have the same usage?"[25] Logistics defines them as if they did by $\sim[(x).\phi x] = \text{df. } (Ex) \sim \phi x$ and insists that since this does not lead to a contradiction the definition is valid.

We are thus led to see that propositions from the finitist point of view may be true (constructible), false, or undecidable. If each of these is a value a propositional function can have, then clearly a three-valued logic is involved such as we mentioned above.[26]

It might be thought that if we take true and false together and lump them under the term decidable, that we have a two-valued logic of decidable, undecidable. However, it can be shown that in such a logic we could construct propositions such that we could not determine whether they were decidable or undecidable. Thus it would appear that we are led to a series: true, false, undecidable, undeterminable (i.e., it cannot be determined whether or not p is undecidable), etc. This means of course that ultimately a logic of n-values is implied (n-indefinite).[27] These distinctions must ultimately be kept and confusions between the symbols for the various negates (e.g., of false, of undecidable, of undeterminable, etc.) must be avoided.[28]

Before proceeding now to the discussion of a number of points of view which are more philosophical in nature concerning mathematics as a whole, it might be useful for the student interested in the logical problems involved to give a brief discussion of modal logics. (Students interested in logic should become acquainted with the *Journal of Symbolic Logic*.)

---

[25] Ambrose, *op. cit.*, p. 198.

[26] The statement made by Miss Ambrose, (p. 317) is based on a confusion of truth-value ranges.

[27] I presented this interpretation in 1934. Cf. Kattsoff, "Postulational Methods I," *Phil. of Science*, Vol. 2 (April, 1935), p. 155; also "Post. Methods II," *ibid.*, Vol. 3, p. 83. For the logic of Probability and Modality, cf. Kattsoff, "Modality and Probability," *Phil. Review*, Vol. 46 (1937), pp. 78–85; Reichenbach, *Wahrscheinlichkeitslehre*, Leyden, 1935, and for a development of order of truth-values, C. G. Hempel, "A Purely Topological Form of Non-Aristotelian Logic," *Journal of Symbolic Logic*, Vol. 2 (1937), pp. 97–112.

[28] For a logic which attempts to make all modal distinctions but yet keep law of contradiction fundamental, cf. H. B. Smith, "Algebra of Propositions," *Phil. of Science*, Vol. 3 (1936), pp. 551–78.

APPENDIX: MODAL LOGICS[29]

Modal logics are best approached through the matrix such as was set up in the chapter on logistics for the values *T* and *F*. Another convention we agree upon is to symbolize the *T* by 1 and *F* by 0. If we do this then the simple matrices for a two-valued logic are:

P	$\sim P$
1	0
0	1

P	Q	$P \vee Q$	$P \rightarrow Q$
1	1	1	1
1	0	1	0
0	1	1	1
0	0	0	1

If we consider $f(x)$ to be a propositional function which is capable of two values 0 and 1, then the possibility arises of having an $f(x)$ of (say) three values, 0, ½, 1—where a possible interpretation of the symbols is 0 = falsity, ½ = possibility, 1 = truth. The problem then arises of determining the structure of such a system. We define the three-valued system in terms of the following matrices:

p	$\sim p$
1	0
½	½
0	1

;

$\rightarrow$ (p \ q)	1	½	0
1	1	½	0
½	1	1	½
0	1	1	1

$\vee$ (p \ q)	1	½	0
1	1	1	1
½	1	½	½
0	1	½	0

It is possible to read the second matrix above as follows: For example, if p is 1 (= certain) and q is ½ (= possible), then $p \rightarrow q$ is ½ (= possible). Notice that the last line indicates that when $p = 0$ (is false) then it implies q certainly, whether q is 1 (= true), ½ (= possible),

[29] Cf. Lewis and Langford, *Symbolic Logic*, New York City, 1932, pp. 213 ff. Zawirski, "Les Logique Nouvelles et le champ de leur application," *Rev. d. Meta. et de Morale*, Vol. 39 (1932), pp. 503 ff. Becker, "Zur Logik der Modalitaten," *Jahrbuch f. Phil. u. Phan. Forsch.*, Vol. 11 (1930), pp. 497 ff. Also, the essay by H. Weyl, "The Ghost of Modality" in *Philosophical Essays in Memory of E. Husserl*, pp. 278 ff., is excellent. Cf. also R. Carnap, *Meaning and Necessity*, Chicago, 1947.

or 0 (=false). This is analogous to the statement that a false proposition, in a two-valued logic, implies anything.

If one keeps in mind the distinction between $\sim p$ and 0 (the former is a *function*, the latter a truth value), then we can introduce a function of p analagous to the truth value ½, i.e., Mp *read* "p is possible." We have another matrix

p	Mp
1	1
½	1
0	0

(The student should now try to verify $p \to . q \to r : \to : q \to . p \to r$ by the matrix method.)

A host of new distinctions arise: Mp, p, and $\sim p$ are all different, $M \sim p$ and $\sim Mp$ are different, etc.

(Verify that $p \to : p \to q . \to q$ is true but that $p \& . p \to q : \to q$ is not true in this logic.)

The system of strict implication given by Professor C. I. Lewis approaches the logic from a different angle.[30]

I. UNDEFINED CONCEPTS

1. Propositions: p, q, r, etc.
2. Negation: $-p$; i.e., *p is false.*
3. Impossibility: $\sim p$; i.e., *p is impossible.*
4. Logical Product: pq or $p \times q$; i.e., p and q are both true.
5. Equivalence: $p = q$.

II. SIMPLEST DEFINITIONS

A. Modalities

(1) p: p is true.

(2) $-p$: p is false.

[30] We follow here Becker's exposition, *op. cit.*, pp. 502 ff. The reader should consult the Lewis and Langford, *Symbolic Logic*, pp. 492 ff., for a comparison between Lewis' system and Becker's.

(3) $\sim p$: p is impossible.

(4) $-\sim p$: it is false that p is impossible, i.e., p is possible.

(5) $\sim - p$: it is impossible that p is false.

(6) $- \sim - p$: p is not necessary, i.e., p is possibly false.

(7) $\sim - \sim p$: p is impossibly possible, i.e., it is impossible that p is possible.

(8) $- \sim - \sim p$: p is not impossibly possible, i.e., it is possible, that p is possible, etc. (The law of advance is not clear. No. 8 is irreducible in Lewis' system, but not in Becker's.)

B. Simple relations between propositions

Definitions:

1.01 Consistency
$p \circ q = - \sim (pq)$ Def.

1.02 Strict Implication
$p < q = \sim (p - q)$ Def.

1.03 Material Implication
$p \rightarrow q = -(p - q)$ Def.

1.04 Strict Logical Sum
$p \wedge q = \sim (-p - q)$ Def.

1.05 Material Logical Sum
$p + q = -(-p - q)$ Def.

1.06 Strict Equivalence
$(p = q) = (p < q)(q < p)$ Def.

1.07 Material Equivalence
$(p = q) = (p \rightarrow q)(q \rightarrow p)$ Def.

The relations called *material* are those of the logistic and formalistic logics. They involve mere negation. The relations called *strict* are those involving modals. The differences between the strict and material relations may be referred back to the readings of the negations. In strict implication $p \rightarrow q$ means that the truth of p and the falsity of q are *logically* impossible together, in material implication the truth of p and the falsity of q are merely *false* together. Strict implication distinguished the logically necessary from the merely true, and the logically impossible from the merely false. 1.01 consistency is not the same concept as is involved in the consistency of a set of postulates.

III. AXIOMS

1.1 $pq < qp$

1.2 $qp < p$

1.3 $p < pp$

1.4 $p(qr) < q(pr)$

1.5 $p < -(-p)$

1.6 $(p<q)(q<r) < (p<r)$

1.7 $\sim p < -p$

1.8 $(p<q) < (\sim q<\sim p)$

Becker adds to this set of axioms:

1.9 $-(\sim p) < \sim(\sim p)$

The set reduces to one of six fundamental modalities [(1)–(6) above]. The following reductions can then be made:[31]

1. $(-\sim)^n p = (\sim\sim)^n p = (\sim)^{2n} p = -\sim p$

$(-\sim)^n p$ means that $-\sim$ occurs n times, i.e., $-\sim-\sim-\sim\ldots-\sim p$, $(\sim)^{2n} p$ means $\sim$ occurs an even number of times. $(n = 1, 2\ldots)$

2. $-(-\sim)^n p = --\sim p = \sim p$

3. $(-\sim)^n - p = -\sim - p$

4. $-(-\sim)^n - p = --\sim - p = \sim - p.$

---

[31] Cf. Becker, *op. cit.*, pp. 508–9.

CHAPTER TWELVE

GODEL'S THEOREM AND FORMAL SYSTEMS

Frequent reference to Gödel's theorem makes it advisable to consider the results known by this name. It is doubly important because of the implications of this famous theorem for the formalist problem.

It is recalled that the formalist approach sets itself two problems: (1) to formalize all mathematics and logic, and (2) to demonstrate the formal consistency of the resultant formalization. In our discussion of von Neumann's paper it was there evident that the problem of consistency depended ultimately upon the problem of decidability—i.e., the problem of deciding whether a given formula is demonstrable (can be shown to be an axiom or the end formula of a schema a and $a \to b$ where a and $a \to b$ are demonstrable) or not (is neither an axiom nor the b of such a schema). Von Neumann had, as we saw, defined consistency in terms of demonstrability.

In the classic Aristotelian logic and in the logistic logic every formula a is either true, i.e., demonstrable, or false, i.e., $\sim a$ is demonstrable. Hence a system is consistent if for every given formula either a or $\sim a$ can be shown to be demonstrable, but not both. Consistency proofs took the form of an attempt to show that for any formula a which can be constructed within a given system and is demonstrable in that system, $\sim a$ is not demonstrable in that system.

The result of the investigations are given in Gödel theorems to which we turn our attention. We can sum up the net result in the words of Carnap: "Everything mathematical can be formalized, but mathematics cannot be exhausted in any one system. . . ."[1] Brouwer was right. Hilbert's second problem, to demonstrate formal consistency, was unsolvable if the consistency proof is to be sought within the system itself. Hilbertian formalism, as originally conceived, is doomed to failure. This is true even though, as Gödel himself pointed out, a consistency proof may be possible in some meta-system or in (perhaps) the "logic of ordinary discourse." But *within* the formal system itself consistency cannot be demonstrated. The first indication that there existed undecidable formulae, i.e., formulae constructible within the system, yet such that neither the formula nor its negation was demon-

[1] R. Carnap, "Die Antinomien und die Unvollstandigkeit der Mathematik"; *Monatshefte für Mathematik u Physik*, Vol. 41 (1934), p. 274.

strable in the system, was given in the now classic paper of Gödel entitled "Uber formal unentscheidbare Sätze der Principia Mathematica und verwandter System" which was published in *Monatshefte für Mathematik und Physik*, Vol. 38, 1931, pp. 173–98.[2]

Gödel recognized at once that this was not a circumstance due to the special nature of *P.M.* but was true of a whole class of formal systems (p. 174). These results were later extended by Church and Rosser among others.[3] In 1934, K. Gödel gave a series of lectures at the Institute for Advanced Study, Princeton, New Jersey, in which he made certain modifications. Notes on these lectures were taken by S. C. Kleene and J. B. Rosser and mimeographed under the title "On Undecidable Propositions of Formal Mathematical Systems." In our discussion we follow the first paper, inserting references to the Princeton lectures where it is deemed necessary.

Gödel's first paper divides into two parts (a) a short sketch of the proof to be offered, and (b) the lengthy process of constructing the exact proof.

Rosser gives the two Gödel theorems as follows:

1. For a formal logic L, satisfying certain conditions (to be stated later), there are undecidable propositions in L; i.e., propositions F such that neither F nor $\sim F$ is provable.

2. For suitable L, the simple consistency of L cannot be proved in L.

The proofs of these theorems are derived by the trick of translating all propositions of L into arithmetical statements by means of a correlation between basic concepts and numbers, and then demonstrating that these arithmetical propositions are undecidable by use of a *reductio* proof. The point of view is entirely formal and all formulae of the system are considered to be finite sequences of fundamental symbols (variables, logical constants, brackets, etc.). Even a proof is defined as a finite sequence of formulae (i.e., of finite sequences of symbols) in which each formula is either an axiom or an immediate consequence of one or more of the preceding formulae. (On Undecidable Propositions, p. 1.)

After the translation of every symbol into a natural number, all formulae and proofs will consist of finite sequences of natural numbers. Since, further, natural numbers are definable in terms of the symbols of *Principia Mathematica*, all metamathematical concepts with which

[2] The reader is referred to Rosser, "An informal exposition of Proofs of Gödel's Theorems and Church's Theorems," *Jour. Symbolic Logic*, Vol. 4, No. 2 (1939), pp. 53–60 for a nontechnical outline of these proofs. Also J. Findlay, "Goedelian Sentences: A Non-Numerical Approach," in *Mind*, July, 1942, pp. 259 ff.

[3] For the literature refer to *The Journal of Symbolic Logic* issues beginning with Vol. 1 to date, especially Vol. 1, No. 3; Vol. 2, No. 3, and Vol. 4, No. 2.

we shall be concerned can be translated into *P.M.* symbols. There fore, such concepts as "formal," "demonstrable," "proof" are so defin able. This means that we can give a formula $F(v)$ from *P.M.* with a free variable v, such that $F(v)$ *when interpreted* means "v is a demonstrable formula." Now, if it be possible so to construct this formula that it can be shown to be undemonstrable, then the theorem will have been demonstrated.

The following procedure outlines a process of constructing such an undecidable proposition in *P.M.* Since we are to translate all propositions into sequences of natural numbers, we are concerned with the construction of a proposition about natural numbers. Any formula with only one free variable, and, more precisely, of the same type as natural numbers, will be called a *class symbol.* (It will be recalled that Russell had defined natural number as a class of classes.)

Imagine the class symbols to be arranged in a series in some way. Let $R(n)$ denote the nth class symbol. (R and class symbol are definable in *P.M.*) Let α be *any* class symbol. Then $[\alpha; n]$ will denote that formula which results from α when we replace any free variable occurring in it by the symbol for the natural number n.

$x = [y; z]$ can also be defined in *P.M.*

We define a class K of natural numbers

(1) $n \epsilon K \equiv \overline{Bew}[R(n); n]$

($Bew\ x$ means "x is a demonstrable formula" and $\overline{Bew\ x}$ means negation of $Bew\ x$.)

Since all the concepts used are definable in *P.M.*, K is also; i.e., there exists a class symbol S such that $[S; n]$ *when interpreted* means that the natural number n belongs to K.

Since S is a class symbol, then $S = R(q)$ for a certain natural number q, i.e., S is the qth class symbol. It is now possible to show that

$[R(q); q]$ is undecidable in *P.M.*

To assume $[R(q); q]$ demonstrable, is to assert that $q \epsilon K$. But if $q \epsilon K$ then it is defined as $\overline{Bew}[R(q);q]$ which means that $[R(q);q]$ is *not demonstrable*, i.e., $[R(q);q]$ is false.

To assume $[R(q);q]$ is false is to assume $q \epsilon K$, i.e., $[R(q);q]$ is demonstrable.

Either assumption thus leads to a contradiction and the proposition cannot be decided.

We now turn to the outline of Gödel's *exact proof.*

Let P be the system obtained when we add Peano's axioms for arithmetic to *P.M.*

Symbols in P

I. Constants:

$\sim$ (not)	$\vee$ (or)	Π (for all)
0 (null)	f (follower of)	( ) (brackets)

II. Variables:

Type 1, (for individuals, i.e., natural numbers including o)
"x_1," "y_1," "z_1," . . .

Type 2, (for classes of individuals)
"x_2," "y_2," "z_2," . . .

Type 3, (for classes of classes of individuals)
"x_3," "y_3," "z_3," . . .

etc.

By a *symbol of Type 1*, we mean a combination of symbols of the form:

$a, fa, ffa, fffa, \ldots$, etc.,

where a is either 0 or a variable of Type 1.

If a is 0 we call such a symbol a *number symbol.*

For $n>1$, we mean by a *symbol of type n*, the same thing as a *variable* of *type n.*

Symbol combinations of the form $a(b)$ where b is a symbol of type n and a is a symbol of type $n+1$, we call *elementary forms.*

The *smallest class* of formulae is that class which contains all elementary formulas and, if it contains a and b, then also $\sim(a)$, $(a) \vee (b)$, $x\Pi(a)$.

Meaningful formulae[4] are the following types:

There are two classes of meaningful formulae.

a) Formulae denoting numbers. This comprises numerical symbols or expressions representing numbers as well as functional expressions which become numerical expressions when numerical expressions are suitably substituted in them. These are called expressions of 1st kind (Exp. I).

b) Formulas denoting propositions: This comprises propositions and propositional functions. These are called expressions of 2nd kind (Exp. II).

[4] This definition is from "On Undecidable Propositions," p. 7.

The exact definition of "meaningful formula" is given inductively as follows:

1) $0, x, y, z, \ldots$ (variables for numbers) are expressions of 1st kind and $p, q, r, \ldots$ (variables for propositions) are expressions of 2nd kind.

2) If A and B are Exp. I, then $A=B$ is Exp. II.

3) If A is Exp. I, then $N(A)$ is Exp. I. ($N(A)$ means "next to A." For example, if A is 0, $N(A)$ is 1.)

4) If A is Exp. I and if f is a variable for a function, then $f(A)$ is Exp. I.

5) If A and B are Exp. II, then $\sim(A)$, $(A) \vee (B)$, $(A) \& (B)$, $(A) \rightarrow (B)$, $(A) \equiv (B)$ are Exp. II.

6) If A is Exp. II and x is a variable for a number, then $\Pi x(A)$ and $\Sigma x(A)$ are Exp. II and $\epsilon x(A)$ is Exp. 1. ($\epsilon x(A)$ is read "the least integer formula $A(x)$ if there is one, otherwise 0.")

7) If A is Exp. II and f is a variable for a function, then $\Pi f(A)$ and $\Sigma f(A)$ are Exp. II.

8) If A is Exp. II and p is a variable for a proposition, then $\Pi p(A)$ and $\Sigma p(A)$ are Exp. II.

To each occurrence of Π, Σ, and ϵ in a meaningful expression A, there corresponds a unique part B of A beginning with Π, Σ, or ϵ of the form $\Pi t(B)$, or $\Sigma t(B)$ or $\epsilon t(B)$, where t is a variable and B is meaningful. This part we call the *scope* of the given Π, Σ, or ϵ in A.

A given occurrence of a variable t in A shall be called *bound*, if it is in the scope of a Π, Σ, or ϵ. Otherwise it is called *free*.[5]

"Subst$a\binom{v}{b}$" indicates the formula resulting from a when we substitute b for v wherever it occurs free.

AXIOMS

I. 1. $\sim(fx_1=0)$ [read: it is false that the follower of x_1 is 0.]

2. $fx_1=fy_1 \supset x_1=y_1$

3. $x_2(0) . x_1\Pi(x_2(x_1) \supset x_2(fx_1)) \supset x_1\Pi(x_2(x_1))$

(Note that in this case the Π follows the variable to which it refers.) (These are some of Peano's axioms.)

[5] "On Undecidable Propositions," p. 8.

II. Formulae resulting from

1. $p \vee p \supset p$
2. $p \supset p \vee q$
3. $p \vee q \supset q \vee p$
4. $(p \supset q) \supset (r \vee p \supset r \vee q)$

when formulae are inserted for p, q, and r.

III. Formulae resulting from

1. $v\Pi(a) \supset \text{Subst}\, a\binom{v}{c}$
2. $v\Pi(b \vee a) \supset b \vee v\Pi(a)$

when we make the following insertions:

any variable for a
any variable for v
a formula in which v does not occur free, for b
a symbol of the same type as v, provided that c does not contain a variable bound in a in a position where v is free.

IV. Formulae resulting from

1. $x_1\Pi(x_2(x_1) \equiv y_2(x_1)) \supset .\, x_2 = y_2$

by raising the type of all variables by the same amount. (This means that a class is determined by its elements.)

CORRELATIONS WITH NATURAL NUMBERS

0 with 1
f with 3
$\sim$ with 5
$\vee$ with 7
Π with 9
(with 11
) with 13

variables of type n, with the numbers of the form p^n (where p is a prime greater than 13).

$n_1n_2n_3 \ldots n_k$ with $2^{n_1}3^{n_2}5^{n_3} \ldots p_k^{n_k}$, where p_k is the k'th prime.

$n_1n_2n_3 \ldots n_k$ is a finite sequence of natural numbers corresponding to the finite sequence of symbols.

We shall denote the number correlated with the basic symbol a by $\phi(a)$. Given $R(a_1a_2 \ldots a_n)$, a class or relation between basic symbols or series of symbols. We correlate with it that class $R'(x_1x_2 \ldots x_n)$ of natural numbers which results from $x_1x_2 \ldots x_n$ when there exist $a_1a_2 \ldots a_n$ such that

$x_i = \phi(a_i)$, $(i = 1, 2, \ldots n)$

and $R(a_1a_2 \ldots a_n)$ holds.

A particular method of defining functions is needed here. This is the *recursive* definition.

A function concerning numbers, say $\phi(x_1x_2\ldots x_n)$ is said to be *recursively defined*[6] from number theoretical functions $\psi(x_1x_2\ldots x_{n-1})$ and $\beta(x_1x_2\ldots x_{n+1})$ if for all $x_2,\ldots x_n, k$

$$\phi(0, x_2\ldots x_n)=\psi(x_2\ldots x_n)$$
$$\phi(k+1, x_2\ldots x_n)=\beta(k, \phi(k, x_2,\ldots x_n), x_2\ldots x_n)$$

A number theoretical function ϕ is said to be a *recursive function* if there exists a finite sequence of number theoretical functions ϕ_1, ϕ_2, $\phi_3\ldots\phi_n$ ending with ϕ and which has the property that every function ϕ_k is either recursively defined from two of the preceding functions or results from any preceding one by substitution or is a constant or the successor function $x+1$.

A relation between natural numbers $R(x_1\ldots x_n)$ is recursive if there exists a recursive function $\phi(x_1\ldots x_n)$ such that for all $x_1, x_2,\ldots x_n$

$$R(x_1\ldots x_n) \sim [\phi(x_1\ldots x_n)=0]$$

(where $\sim$ denotes Hilbert's $\leftrightarrows$).

Definitions of metamathematical classes and relations of classes.

1. $x|y \equiv (Ez)[z\leqq x \& x=y.z]$

Read: "x is divisible by y," means that there is a z such that z is less than or equal to x and x is the product of y and z.

2. $\mathrm{Prim}(x) \equiv (\overline{Ez})[z\leqq x \& z\neq 1 \& z\neq x \& x|z] \& x>1$

Read: "x is a prime" means that there does not exist a z such that z is less than or equal to x and z is not 1 or x and x is divisible by z—provided x is greater than 1.

3. $o\mathrm{Pr}x \equiv 0$
$(n+1)\mathrm{Pr}x \equiv \epsilon y[y\leqq x \& \mathrm{Prim}(y) \& x|y \& y>n\mathrm{Pr}x]$
("$n\mathrm{Pr}x$" means the nth prime in x.)

4. $0! \equiv 1$
$(n+1)! \equiv (n+1)n!$

5. $\mathrm{Pr}(0) \equiv 0$
$\mathrm{Pr}(n+1) \equiv \epsilon y[y\leqq \{\mathrm{Pr}(n)\}!+1 \& \mathrm{Prim}(y) \& y>\mathrm{Pr}(n)]$
($\mathrm{Pr}n$ is the nth prime.)

---

[6] Cf. Gödel, "On Undecidable Propositions," pp. 2 ff.; Church, "An Unsolvable Problem of Elementary Number Theory," *American Journal of Mathematics*, Vol. 58, No. 2 (April, 1936), pp. 350 ff.

6. $n\mathrm{Gl}x \equiv \epsilon y[y \leqq x \& x|(n\mathrm{Pr}x)^y \& \overline{x|(n\mathrm{Pr}x)^{y+1}}]$
 $n\mathrm{Gl}x$ is the nth element (Glied) of the sequence of numbers correlated to the number x. This means the sequence of exponents in $x = 2^{k_1}3^{k_2}5^{k_3}\ldots$

7. $L(x) \equiv \epsilon y[y \leqq x \& y\mathrm{Pr}x > 0 \& (y+1)\mathrm{Pr}x = 0]$
 $L(x)$ is the length of the series correlated to x

8. $x*y \equiv \epsilon z\{z \leqq [\mathrm{Pr}(L(x)+L(y))]^{x+y} \& (n)[n \leqq L(x) \rightarrow$
 $n\mathrm{Gl}z = n\mathrm{Gl}x] \& (n)[0 < n \leqq L(y) \rightarrow$
 $(n+L(x))\mathrm{Gl}z = n\mathrm{Gl}y]\}$
 $x*y$ is the operation of "joining together" two finite number series. For example, if $x = 2^{k_1}3^{k_2}\ldots p_r^{k_r}$ and $y = 2^{l_1}3^{l_2}\ldots p_s^{l_s}$, then $x*y = 2^{k_1}\ldots p_r^{k_r}p_{r+1}^{l_r}\ldots p_{r+s}^{l_s}$

9. $R(x) \equiv 2^x$
 $R(x)$ corresponds to the number series consisting only of the number x. (for $x > 0$.)

10. $E(x) \equiv R(11)*x*R(13)$
 $E(x)$ corresponds to operation of "enclosing in brackets," since 11 and 13 are correlated to "("and")." Note that the difference between "enclosing in brackets" and the existential quantifier lies in the fact that the E is outside the parentheses for "enclosing in brackets" and inside for the existential quantifier.

11. $n\mathrm{Var}x \equiv (Ez)[13 < z \leqq x \& \mathrm{Prim}(z) \& x = z^n] \& n \neq 0$
 x is a *variable of type n*.

12. $\mathrm{Var}x \equiv (En)[n \leqq x \& n\mathrm{Var}x]$
 x is a *variable*.

13. $\mathrm{Neg}(x) \equiv R(5)*E(x)$
 $\mathrm{Neg}(x)$ is *negation* of x.

14. $x\mathrm{Dis}y \equiv E(x)*R(7)*E(y)$
 $x\mathrm{Dis}y$ is the *disjunction* of x and y.

15. $x\mathrm{Gen}y \equiv R(x)*R(9)*E(y)$
 $x\mathrm{Gen}y$ is the *generalization* of y by means of variable x.

16. $0\mathcal{N}x \equiv x$
 $(n+1)\mathcal{N}x \equiv R(3)*n\mathcal{N}x$
 $n\mathcal{N}x$ corresponds to operation "putting f before x n-times."

17. $Z(n) \equiv n\mathcal{N}[R(1)]$
$Z(n)$ is the *number symbol* for the number n.

18. $\mathrm{Typ}_1(x) \equiv (Em,n)\{m,n \leqq x \& [m=1 \vee 1\mathrm{Var}m] \& x = n\mathcal{N}[R(m)]\}$
x is a *symbol of first type.*

19. $\mathrm{Typ}_n(x) \equiv [n=1 \& \mathrm{Typ}_1(x)] \vee [n>1 \& (Ev)\{v \leqq x \& n\mathrm{Var}v \& x = R(v)\}]$
x is a *symbol of type n.*

20. $\mathrm{Elf}(x) \equiv (Ey,z,n)[y,z,n \leqq x \& \mathrm{Typ}_n(y) \& \mathrm{Typ}_{n+1}(z) \& x = z * E(y)]$
x is an *elementary formula.*

21. $\mathrm{Op}(xyz) \equiv x = \mathrm{Neg}(y) \vee x = y\mathrm{Dis}z \vee (Ev)[v \leqq x \& \mathrm{Var}(v) \& x = v\mathrm{Gen}y]$

22. $FR(x) \equiv (n)\{0<n \leqq L(x) \rightarrow \mathrm{Elf}(n\mathrm{Gl}x) \vee (Ep,q)[0<p,q<n \& \mathrm{Op}(n\mathrm{Gl}x, p\mathrm{Gl}x, q\mathrm{Gl}x)]\} \& L(x)>0$
x is a *series of formulas,* each of which is either an elementary formula or is derived from the preceding form by operations of Neg., Dis., Gen.

23. $\mathrm{Form}(x) \equiv (En)\{n \leqq (\mathrm{Pr}[L(x)^2])^{x.[L(x)]^2} \& FR(n) \& x = [L(n)]\mathrm{Gl}n$
x is a *formula.*

24. $v\mathrm{Geb}n,x \equiv \mathrm{Var}(v) \& \mathrm{Form}(x) \& (Ea,b,c)[a,b,c \leqq x \& x = a*(v\mathrm{Gen}b)*c \& \mathrm{Form}(b) \& L(a)+1 \leqq n \leqq L(a)+L(v\mathrm{Gen}b)]$

The variable v is *bound* to the nth place in x.

25. $v\mathrm{Fr}n,x \equiv \mathrm{Var}(v) \& \mathrm{Form}(x) \& v = n\mathrm{Gl}x \& n \leqq L(x) \& \overline{v\mathrm{Geb}n,x}$
Variable v is *free* in nth place in x.

26. $v\mathrm{Fr}x \equiv (En)[n \leqq L(x) \& v\mathrm{Fr}n,x]$
v occurs in x as a *free* variable.

27. $\mathrm{Su}x\binom{n}{y} \equiv \epsilon z\{z \leqq [\mathrm{Pr}(L(x)+L(y))]^{x+y} \& [(Eu,v)\ u,v \leqq x \& x = u*R(n\mathrm{Gl}x)v \& z = u*y*v \& n = L(u)+1]\}$

$\mathrm{Su}x\binom{n}{y}$ results from x if we substitute y for the nth element of x $(0<n \leqq L(x))$

28. $0\mathrm{St}v,x \equiv \epsilon n\{n \leqq L(x) \& v\mathrm{Fr}n,x \& (\overline{Ep})[n<p \leqq L(x) \& v\mathrm{Fr}p,x]\}$
$(k+1)\mathrm{St}v,x \equiv \epsilon n\{n<k\mathrm{St}v,x \& v\mathrm{Fr}n,x \& (\overline{Ep})[n<p<k\mathrm{St}v,x \& v\mathrm{Fr}p,x]\}$

kStv,x is the $k+1$st place (Stelle) in x (counted from the end of formula x) in which v is free in x (if no such place exists it is 0).

29. $A(v,x) \equiv \epsilon n\{n \leqq L(x) \& n\mathrm{St}v,x = 0\}$
$A(v,x)$ is the number of places in which v is free in x.

30. $\mathrm{Sb}_0\left(x\begin{smallmatrix}v\\y\end{smallmatrix}\right) \equiv x$

$\mathrm{Sb}_{k+1}\left(x\begin{smallmatrix}v\\y\end{smallmatrix}\right) \equiv \mathrm{Su}\left\{\mathrm{Sb}_k\left(x\begin{smallmatrix}v\\y\end{smallmatrix}\right)\right\}\left(\begin{smallmatrix}k\mathrm{St}v,y\\y\end{smallmatrix}\right)$

31. $\mathrm{Sb}\left(x\begin{smallmatrix}v\\y\end{smallmatrix}\right) \equiv \mathrm{Sb}_{A(v,x)}\left(x\begin{smallmatrix}v\\y\end{smallmatrix}\right)$

$\mathrm{Sb}\left(x\begin{smallmatrix}v\\y\end{smallmatrix}\right) \equiv$ is the concept Subst.$a\left(\begin{smallmatrix}v\\b\end{smallmatrix}\right)$.

32. $x\mathrm{Imp}y \equiv [\mathrm{Neg}(x)]\mathrm{Dis}y$
$x\mathrm{Con}y \equiv \mathrm{Neg}\{[\mathrm{Neg}(x)]\mathrm{Dis}[\mathrm{Neg}(y)]\}$
$x\mathrm{Aeq}y \equiv (x\mathrm{Imp}y)\mathrm{Con}(y\mathrm{Imp}x)$
$v\mathrm{Ex}y \equiv \mathrm{Neg}\{v\mathrm{Gen}[\mathrm{Neg}(y)]\}$
This represents $\sim\Pi v(\sim y)$, i.e., the negation of a generalization.

33. $n\mathrm{Th}x \equiv \epsilon y\{y \leqq x^{(x^n)} \& (k)[k \leqq L(x) \rightarrow (k\mathrm{Gl}x \leqq 13 \& k\mathrm{Gl}y = k\mathrm{Gl}x) \vee (k\mathrm{Gl}x > 13 \& k\mathrm{Gl}y = k\mathrm{Gl}x.[1\mathrm{Pr}(k\mathrm{Gl}x)]^n)]\}$
nThx is the nth *raising* of the *type* of x.

To axioms I, 1–3 there correspond three definite numbers indicated as z_1, z_2, z_3 and defined as

34. $Z\text{-Ax}(x) \equiv (x = z_1 \vee x = z_2 \vee x = z_3)$

35. $A_1\text{-Ax}(x) \equiv (Ey)[y \leqq x \& \mathrm{Form}(y) \& x = (y\mathrm{Dis}y)\mathrm{Imp}y]$
x is a *formula* resulting from substitution in the schema II, 1.

36. $A\text{-Ax}(x) \equiv A_1\text{-Ax}(x) \vee A_2\text{-Ax}(x) \vee A_3\text{-Ax}(x) \vee A_4\text{-Ax}(x)$
x is a *formula* resulting by substitution in an appropriate axiom.

37. $Q(z,y,v) \equiv \overline{(En,m,w)}[n \leqq L(y) \& m = L(z) \;\&\; w \leqq z \;\&\; w = m\mathrm{Gl}z \;\&\; w\mathrm{Geb}n,y \& v\mathrm{Fr}n,y]$
z contains no variable bound in y in a place where v is free.

38. $L_1\text{-Ax}(x) \equiv (Ev,y,z,n)\{v,y,z,n \leqq x \;\&\; n\mathrm{Var}v \;\&\; \mathrm{Typ}_n(z)$

$\& \mathrm{Form}y \;\&\; Q(z,y,v) \;\&\; x = (v\mathrm{Gen}y)\mathrm{Imp}[\mathrm{Sb}\left(y\begin{smallmatrix}v\\z\end{smallmatrix}\right)]\}$

x is a formula resulting from III, 1 by substitution.

39. $L_2\text{-Ax}(x) \equiv (Ev,q,p)\{v,q,p \leqq x \,\&\, \text{Var}(v) \,\&\, \text{Form}(p) \,\&\, \overline{v\text{Fr}p} \,\&\, \text{Form}(q) \,\&\, x = [v\text{Gen}(p\text{Dis}q)]\text{Imp}[p\text{Dis}(v\text{Gen}q)]\}$

    x is a formula resulting by substitution from III, 2.

40. $R\text{-Ax}(x) \equiv (Eu,v,y,n)[u,v,y,n \leqq x \,\&\, n\text{Var}v \,\&\, (n+1)\text{Var}u \,\&\, \overline{u\text{Fr}y} \,\&\, \text{Form}(y) \,\&\, x = uEx\{v\text{Gen}[[R(u) * E(R(v))]\text{Aeq}y]\}]$

    x is a formula resulting from IV, 1 by substitution.

A definite number z_4 corresponds to axiom V, 1 and we define

41. $M\text{-Ax}(x) \equiv (En)[n \leqq x \,\&\, x = n\text{Th}z_4]$

42. $\text{Ax}(x) \equiv Z\text{-Ax}(x) \vee A\text{-Ax}(x) \vee L_1\text{-Ax}(x) \vee L_2\text{-}Ax(x) \vee R\text{-Ax}(x) \vee M\text{-Ax}(x)$

    x is an *axiom*.

43. $\text{Fl}(xyz) \equiv y = z\text{Imp}x \vee (Ev)[v \leqq x \,\&\, \text{Var}(v) \,\&\, x = v\text{Gen}y]$

    x is an immediate consequence (Folge) of y and z.

44. $Bw(x) \equiv (n)\{0 < n \leqq L(x) \rightarrow \text{Ax}(n\text{Gl}x) \vee (Ep,q)[0 < p,q, < n \,\&\, \text{Fl}(n\text{Gl}x, p\text{Gl}x, q\text{Gl}x)]\} \,\&\, L(x) > 0$

    x is a *proof-schema* (a finite sequence of formulas each of which is either an axiom or an immediate consequence of two of its predecessors).

45. $xBy \equiv Bw(x) \,\&\, [L(x)]\text{Gl}x = y$

    x is a *proof* for y.

46. $Bew(x) \equiv (Ey)yBx$

    x is a *demonstrable formula*.

With these we are ready to reach the goal of these investigations.

Let α be any class of formulae.

Let $\text{Flg}(\alpha)$ be the least class of formulae which contains all formulae of α and all axioms and is closed with reference to the relation "immediate consequence."

α is ω-consistent means there exists no class symbol a such that

$$(n)[Sb\left(a\begin{smallmatrix}v\\ Z_{(n)}\end{smallmatrix}\right)\epsilon\text{Flg}(\alpha)] \,\&\, [\text{Neg}(v\text{Gen}a)]\epsilon\text{Flg}(\alpha)$$

(When interpreted this means: no property F exists for which both all formulae $F(n)$ belong to α (i.e., are demonstrable) and $(Ex)\sim F(x)$ belongs to α (i.e., is demonstrable).)

It is possible now to prove

Proposition VI. For every ω-consistent recursive class α of formulae there exist recursive class symbols r, such that neither vGen$r$ nor Neg (vGen$r$) belongs to Flg(α).
(We omit the proof here.)

From VI follow other results.

A relation (class) is called *arithmetical* if it can be defined by means of the concepts $+$ and $.$ and the logical constants $\vee$, —, or $\sim$, (x), $=$ where (x) and $=$ are relative to natural numbers only. For example, "greater than," "congruent modulo n."

Then

VII. Every recursive relation is arithmetic.

VIII. In every ω-consistent system resulting from P (i.e., the system obtained when we add Peano's axioms for arithmetic to Principia Mathematica) by adding a recursively definable class of axioms, there exist undecidable arithmetic propositions (p. 193).

IX. In all such systems there are undecidable problems of the restricted functional calculus (i.e., formulas of the restricted functional calculus for which neither their universal validity nor the existence of a negative example is demonstrable. *See* p. 193).

This rests on

X. Every problem of the form $(x)F(x)$ (F recursive) can be reduced to a question about the existence of an application of a formula of the restricted functional calculus (p. 194).
The restricted functional calculus is composed of all formulas which can be constructed from :—, $\vee$, (x), $=$; $x, y \ldots$, $F(x)$, $G(x,y)$, $H(x,y,z)$,—.

Furthermore

XI. If α is any recursive consistent class of formulae then: The propositional formula asserting that α is consistent is not α-demonstrable; in particular if P is consistent, then the consistency of P is undemonstrable in P (p. 196).

Consistency is defined as:

$$\text{Wid}(\alpha) \equiv (Ex)[\text{Form}(x) \;\&\; \overline{Bew_{\alpha}(x)}]$$

In concluding this outline of Gödel's work, we must again refer to an idea of the intuitionists that mathematics can never be exhausted in any symbolic system. Gödel pointed out in 1932[7] that if we extend P by adding to it variables for classes of higher type than exist in it, and add the axiom of extentionality we get another system P' in which we can demonstrate the ω-consistency of P. But the Gödel theorem can now be applied to P'. We are thus led to a sequence of such systems P_1 such that we can demonstrate the ω-consistency of any P_j in P_{j+1}.

What is the significance of Gödel's theorem? From a purely formalistic point of view, as we have said, it shows that the problem of finding a general method of consistency proof, is unsolvable. In addition, it seems to indicate that under the conditions stated, it is impossible to formalize a logic completely, since each formal system requires a wider system if it is to demonstrate all its formulae. As Church points out (*Mathematical Logic*, pp. 110–12), the conditions are such that all the usual systems of symbolic logic as well as every conceivable satisfactory system satisfy them.

The theorem shows that in the consistency proof of any system of symbolic logic, "the intuitive argument must involve some principle which is incapable of formalization within the system in question." This implies at least the existence of some intuitive logic. "But," goes on Church, "perhaps more important is the inference which may apparently be drawn from Gödel's theorem that no one system of formal logic can embrace all forms of reasoning which are *correct*" (*M. L.*, p. 111). Church offers an alternative interpretation of Gödel's theorem which depends on allowing an infinite set of meanings for the universal quantifier, but this would merely change the difficulty to another place.

If logic is viewed merely as a set of symbols manipulated according to specified rules, then Gödel's theorem means that there must be an infinite regress of such rules, if the logic is to be completely described. The existence of such an infinite regress indicates that at any step the system of rules (i.e., the logic) is incomplete. It does not invalidate the use of the logic at any given stage. If we accept the Aristotelian notion that an infinite series of steps invalidates the series, then clearly there can be no formal logic in the sense that Hilbert had in mind. But the very existence of the intuitive logic needed to build up the formal logic indicates that logic is more than a set of symbols. An intuitive logic, whatever else it may imply, does seem to refer to more than a game with symbols.

Another aspect needs to be indicated. If, as is agreed, the formal systems can be used to systematize fields of knowledge, and if the Gödel theorem applies to such applications, it would mean that no field of knowledge so systematized can ever be complete. At any stage there

[7] K. Gödel, "Uber Vollständigkeit und Widerspruchsfreiheit," *Ergebnisse eines Mathematischen Kolloquiums*, Heft 3, 1932, pp. 12–13.

will always be questions formulable in the system that need a wider system for their solution. Since it is the contention of the author of this book that logic and mathematics do tell us something about reality, the Gödel theorem tells us that there can never be a complete and final theory of reality, i.e., metaphysics. This does not lead to mysticism, but to an evolutionary scheme of reality.

CHAPTER THIRTEEN

MANNOURY'S SIGNIFIC POINT OF VIEW

A RETURN TO A PSYCHOLOGISM

The two chief emphases in intuitionism were upon (1) constructability, (2) the relation of mathematics to the thinking subject. (These ideas are more or less adopted by the formalists whose chief stress seems to be upon *operations*.) Both of these ideas have been given a more general application. The second, especially, if incautiously developed, would seem to lead to a subjectivism of Kantian type. Certainly it can lead to some form of psychologism.

Brouwer himself had tried to avoid subjectivism in the way we described above. If, then, mathematics was related to the thought of the investigator but was not a mere creation of that thought, it may be in some way an appearance or phenomenon of *life* itself. Such is the fundamental thesis of Mannoury.[1]

If this is so, then it ought to be investigated with a view to its relation to other life-phenomena. "Mathematics" therefore divides into two parts: (1) In the narrower sense it means the words and symbols through which it manifests itself. We shall call this the *mathematical form of* appearance or formalistic mathematics. (2) In the wider sense it is the system of mental associations which is the basis of the formalistic mathematics. We call this the mathematical *thought form* or *intuitionist* mathematics.

From this point of view the fundamental question becomes: "Why, by what means, and to what extent is mathematics *pursued* and practiced?" If mathematics is a phenomenon of life (Lebenserscheinung), then the individual mathematician is only one illustration of the appearance of mathematics. Since mathematics makes its appearance through verbal forms, we must first apply our question to verbal forms and their psychological significance.

We shall mean by a *verbal act* (Sprachakt) any action through which living beings attempt to influence the conduct or life behavior of other living beings. By the (significs) meaning of a verbal act

[1] The discussion here is based on Mannoury, "Die signifischen Grundlagen der Mathematik," *Erkenntnis*, Vol. 4, 1934, pp. 288–309 and pp. 317–45. Cf. also Bentley, *Linguistic Analysis of Mathematics*, Bloomington, 1930.

(which may be oral or written or mimicry or by gesticulation) we mean the comprehension of the associations which relate this act to definite mental complexes in the mind of the person in question. In an extended sense these are to be called "speaker" and "hearer."[2] Significs is thus much wider than semantics. A dictionary word is not a verbal act but a form common to many verbal acts.

We define *significs* as the theory of the mental associations which lie behind human verbal acts but exclusive of the sciences of language in the narrower sense (semantology, etymology, linguistics, and philology.[3]

By *signific methods* we mean the investigation of these associations empirically.

Every verbal act is purely subjective since its purpose is to influence other living creatures, but the extent of the satisfaction caused in the speaker varies greatly and enables us to divide verbal acts into two classes. This satisfaction is in many cases determined by the choice of the hearer. Where this is the case we call such verbal acts *proper expressions of the will* (eigentliche Willensausserungen). Where the satisfaction depends, on the whole, on the choice of the hearer but is essentially conditioned by the completion of the verbal act itself we have *communicative or indicative verbal acts.* An example of this is a simple description of an object or occurrence.[4]

If we consider the meaning element involved in a verbal act, we can more readily comprehend the difference in amount of satisfaction of the expectations of the speaker. If we consider the different types of mental complexes which are involved in a given verbal act, we can easily distinguish between invariable and unaffected images and memories on the one hand, and the continuously increasing and diminishing emotional desires and expectations on the other. We define the *indicative meaning elements* of a verbal act as the effective associations in the mind of the speaker or hearer including memory images originating from the experience of external objects or events, or imaginative images constructed from them. *Emotional meaning elements* are all other kinds, especially the associations relating to feelings and impulses.

There are various levels of language, which can be denoted as follows:[5]

a) Basic language (Grundsprache). In this language every word or group of words relates directly to experience. (This is probably what the logical positivists call *protocol* language.[6])

---

[2] *Ibid.*, pp. 289–90.
[3] *Ibid.*, p. 290.
[4] *Ibid.*, p. 291.
[5] *Ibid.*, pp. 296–97.
[6] Cf. Carnap, *Philosophy and Logical Syntax.*

b) Language of subjective connotation (Stimmungssprache). Syntactic relations are clearly experienced but not in a durable fashion. The words act both directly and by means of the memory of other words on the mind of the auditor. To this level belongs the subject-predicate form of language as it is found in occidental society, but only in so far as it restricts itself to the formulation of personal experiences and correspondences (language of the people, of poetry, etc.).

c) Communication language. Syntactic relations have become of paramount importance so that the words almost never aim at an independent effect, and every departure from the traditional grouping of opposites (as white-black, good-bad, freedom-necessity, yes-no) is felt to be improper and confusing.

d) Scientific language. Syntactic relations which depend, at least in great part, on express agreement or condition. This is the language of law and orders (or directions) of financial relations, of technics and science in the narrower sense.

e) Symbolic language. In this are included those logical systems which are based exclusively on pre-established forms of substitution (axioms, postulates, etc.) by means of the symbols used. Here are included mathematical logic and those parts of mathematics put into pasigraphic form or which can be put into such form.

The gap between psychological and physical terminology can be reduced to the ordinary distinction between subject and object, between the "I" and the "It." This distinction is so familiar to us that we can almost speak of a dual language, i.e., an *I-now*-terminology on the one hand and an *It*-terminology on the other. In the "*I-now*-terminology" the indicative meaning elements of the words depend on the person who is speaking, while in the "*It*-terminology" this is not the case. "Berlin," "Paris" whether enumerated by John or Robert denote the same cities, but "I," "here," "now" vary their indicative contents with any variation in speaker.

The ordinary rules of logic apply in the *I-now*-terminology only so long as the speaker does not change. From "I live in *A*" and "I am a *C*" can be deduced "A *C* lives in *A*" only if the two "*I*'s" are identical. In most cases, this distinction leads to no misunderstandings. But there are cases where difficulties do arise. For example, the question "Does the world have a purpose?" If, however, we recollect that "purpose" (Zweck) belongs to the *I-now*-terminology and "world" (Welt) to the *It*-terminology then the question "Does the world have a purpose?" (apart from the affects or impulses associated with this question in the speaker) is as simple a question as if one wanted to know how "horse" is to be written in French, or "cheval" in English. In the *It*-

terminology the word *purpose* does not occur without the indication of some agent. But we cannot conclude that there are no horses in France or that "our entire existence is a purposeless game." On the contrary *every* act of ours aims at something.

We now turn to Mannoury's analysis of mathematics.[7]

What does $\frac{d\sin x}{dx} = \cos x$ aim at? Such a question can be answered just as little as the question "What does *yes* or *no* aim at?" or the question "What does 1 or *m* mean?" A mathematical formula, apart from mere study purposes, is only expressed or written down when it is to be applied to some experiential datum. The purpose of this application is not determined by the formula but by the relation of the person speaking to the object of his calculation. The formula, or more generally, the more or less "pure" mathematics, is only the form of appearance of the transfer from purpose to means.

From this it follows that the completely formalized or pure mathematical language cannot be a means for the transfer of the will and therefore it contains neither emotional nor indicative meaning elements. This does not mean that the speaker who uses a formal language has no purpose, but rather that the associations which connect mental images with the verbal structures are variable in character and do not have the stability which is present in other cases and which enables us to speak of the signific meaning of a word. In a general sense such a signific meaning must be denied mathematics proper. The verbal acts pertaining to mathematics have only an incidental signific meaning.

This is not, however, an absolute distinction between mathematical and nonmathematical languages, but a *gradual* distinction. All levels of language are more or less formalized in that there occur memory-verbal-images which are related to other memory word images in a definite fashion.

The first step in the formalization of language is taken as soon as the first opportunity for a future language of communication has arisen, i.e., as soon as the law abiding recurrence of the same sounds or sound complexes correlated with definite situations has been noticed, we must admit associations of the verbal act with the memory images of the pertinent sounds on the one hand, and with the memory images of the corresponding situations on the other. The road to the formalization of language has been constructed with this. The next stage in the formalization is in the appearance of relation between *different* sound complexes which are independent of the accompanying situation, so that we can speak of a primitive word-image net of associations. At the same time the movements and utterances of the child are co-

[7] Mannoury, *ibid.*, pp. 317 ff.

ordinated with visual and other sensations and experiences and a space-time order arises which at once develops into a (physical) net of associations that enables the person to move his bodily parts more purposefully—to grasp or reject objects, etc.[8]

At this stage one can hardly speak of indicative meaning elements of the verbal act. The language of the child is primarily emotional and volitional.

During the course of years two nets of association—the *physical and logical subconscious complex*—develop, for the most part independently, into an almost complete structure which is in many respects similar in different individuals. This similarity is what makes communication between men possible. But no two people see the world about them from the same point of view and no two people have the same vocabulary and linguistic habits.

Let us consider the physical subcomplex.

Natural science is not a psychic complex but the scientific knowledge of physicist X constitutes a psychic complex which is composed of two parts of different natures. In so far as physicist X can communicate his science by means of words (in lectures or articles) then his knowledge of physics is a partial-complex of his logical subcomplex. But if X is a real physicist—not a mere book-scholar who repeats a memorized lesson uncomprehendingly—then his expert knowledge embraces, in addition to the word-image net of associations, also a net of experiential associations which ties the memory images of the physical experiments together, and *this* net belongs obviously to his physical subcomplex.

As for the concept "necessity in nature" which gives rise to the concept of the various natural laws, these are only expressions for regularities which the associations between remembered expectations and experiences disclose (the content of experience or indicative meaning of natural law) on the one hand, and on the other for the corresponding regularities in the associations between experience and expectation (emotional-volitional meaning of natural law). In the last analysis it is impossible for natural laws to have an objective content in the strictest sense of the word, i.e., a content independent of men or of human life phenomena, because they are ultimately psychological laws. But they are however of great significance for the conduct of our lives since their knowledge makes easier the leap "from end to means."

Indeed there are distinctions of stability and generality in these associations. For example, the formula: "Friday is an unlucky day" expresses a regularity in the relation between definite expectations (bad luck) and definite sensations (of the page on the calendar) which

[8] A behavioristic theory of signs has been developed by C. W. Morris, *Signs, Language, and Behavior*, N. Y., 1946. The developing psychologism is evident in this book, which defines signs in terms of the behavior evoked.

occurs for many of our contemporaries. In this respect the case is the same as with the law of falling bodies except that the regularity expressed by the law of falling bodies occurs more often.

The differences in the stability of these associations are expressed in daily life by the predicates, *necessity of nature*, *probable*, *improbable*, *absolutely*, *impossible*, etc.

The relation between the logical and physical subcomplexes will be clear if we return to the Physicist X as he concerns himself not with experiment but with theory, and develops the consequences of a hypothesis in order to test these consequences by experiment. The deductions and calculations which he writes down constitute a partial formalization of his technical terminology. What he is doing may be termed *synthetic significs.* X is assisted in his work by his linguistic habits of thought, and the mathematical formulae he uses constitute a partial group of his habit of thought. It must be kept in mind then that the logical justification of his conclusions is not identical with the way these conclusions arise in the subconscious of X.

Essentially the same process takes place when X assigns part of the synthesis to others, i.e., to those whom we call mathematicians. These men follow in their work their own linguistic mental habits. But the associations which bridge the gap between the logical and physical subcomplexes need play no part when we speak of "pure" or "formal" mathematics.

If now we consider the set of associations which are related to the language itself, i.e., to auditory, visual, or motor verbal images, we see here also numerous regularities—not only in the linguistic habits of individuals but also in entire peoples and groups of people. We mean those regularities which belong to the sphere of logic: subject-predicate relations, forms of opposition and negation, the relation of conjunction to disjunction, etc.

But we must remark that these regularities are in general less stable than those which our physical subcomplex shows. *Laws of language are not laws of nature and logic is not a natural science.* But these regularities are of the same kind considered, psychologically, as those which are expressed as natural laws: the agreement of recalled expectations and experiences on the one hand and the relating of experience and expectation on the other. This indicates that the investigation of both complexes and the life phenomena ruled by them follows the same principles, and the work of physicist, logician, and significist coincide essentially. When Aristotle, e.g., tries to enumerate the logical figures, his work is to be viewed as a synthetic formalization of his experience acquired by analyzing the verbal habits of his contemporaries.

If the question of the significance of mathematics proper from the signific point of view is raised, then a distinction with respect to the physical and logical sciences is unavoidable. The mental work of the most isolated mathematician consists of linguistic acts. Therefore it can and must be tested in its indicative and emotional meaning elements. It is clearly evident that even this type of mental work is conditioned by memory and expectation associations, as is every other type. Only here the "experiential content" of the mathematical theorem or proof consists in the knowledge common to the author and reader. But we must not lose sight of the difference between a formal mathematical and formal physical linguistic act.

Among the many common technical expressions in formal mathematics whose form agrees with definite fundamental words of ordinary language we have, on the one hand, the symbols for addition and subtraction and, on the other, the words for conjunction and negation: $a+b$ is not read in mathematics as *a and b* nor is $a-b$ read there as *a but not b*. But if we consider the intermediate stages through which the student passes in going from elementary to secondary school we can easily distinguish the transfer of meaning to which we refer. Yet the signific difference between the first and last stages of this transfer is extremely significant. For the formalist, the symbol $+$ is so closely bound up with the formal laws of addition, $a+b=b+a$ and $a+(b+c) = (a+b)+c$, that every time the formalist uses the $+$ he follows these laws.

When we speak of "red *and* yellow," "France *and* Germany," "today *and* tomorrow," the corresponding memory images are perhaps fairly distinguishable and stable. The memory images, in other words, often occur separately and also connected in consciousness even without using the same words. But when we speak of "hale and hearty," "safe and sound," etc., the indicative meaning elements are often much weaker and less stable.

But even in daily life formal laws are obeyed when the conjunction particle is used. These laws agree to a certain extent with the associative and commutative laws of the theory of operations. But the speaker is not always aware of this fact and the use of these laws has much less significance than with the formalist. The order of the "summanda" is in daily life irrelevant but is conditioned by usage or accompanying circumstances. For example, if one speaks of "Mrs. and Mr." instead of "Mr. and Mrs.," the speaker almost certainly has a definite purpose in mind.

So far as negation is concerned the matter is more involved. In daily life when negatives are used, *emotional* meaning elements are prominent. Where we are concerned with opposites (*not* big, *not* permitted, etc.) we have two more or less definite indicative meaning elements present which are united by the verbal use of disjunction (big *or* small, permitted *or* forbidden, clean *or* dirty, etc.). The volitional value of the verbal act expressed in negative form is frequently only slightly differentiated from its corresponding positive form. In the (nonformal) contradictory, however, (that is *im*possible, etc.) either *no* definite disjunction occurs or the attention is not on the antithesis. This brings out the emotional meaning elements which can be experienced and which have the character of a *restriction* or *rejection*. We thus distinguish between a *negative of exclusion* and a *negation of choice*. This emphasizes also an important formal difference between the two forms of negation. The law of excluded middle is followed consistently in the negation of exclusion, but is not followed or at least is followed to a much less degree in negations of choice.

We set up the possibilities and illustrations as follows:

a) double negation of choice = assertion
For example, "The opposite of big city is little city, and *conversely*."

b) exclusion of a negation of choice = assertion *or* "excluded middle"

For example, "What is no small city, can be a big city, but may also be something totally different."

c) negation of choice of an exclusion: does not occur

d) double negation of exclusion = assertion

For example, "If the possibility that *no* big city was meant is *excluded*, then a big city must have been meant."

This distinction is of fundamental importance for the signific investigation of mathematical thought forms, because in the languages of our society we have developed a whole group of expressions and linguistic forms from the negation of exclusion. These may be denoted as the linguistic forms of *universality* and are related to the negation of exclusion by the formulae "*a* or non-*a* = everything" and "*a and* non-*a* = nothing." Other concepts belonging to this same linguistic form (as "infinite," "eternal," "never," "certain," "I," "empty") are to be referred back to these two formulae more or less directly. It is of interest to note that the concepts of *infinity* and of the *null class* at the foundations of the theory of aggregates are concepts of this sort.

In everyday life we feel for the most part a sharp division between those representations pertaining to the physical subcomplex which are most closely related to our immediate expectations and all other representations. The former constitutes a subgroup of our physical subcomplex which we shall denote as our "concept of objective reality or the physical environment" (briefly "circumcomplex"). In practical normal life the boundaries of this subcomplex are rarely destroyed (aside from dreams, hallucinations, and the like).

The (passive) elements immediately connected with this circum-complex and the impulses which accompany them (as active elements) are to be considered as a whole, which is to be described or defined as "self-consciousness" or "concept of subjective reality." We shall denote this as the "*I*-complex." It is to be noted that the circumcomplex and the *I*-complex together constitute only a very small part of the content of our minds.

Let us now compare a simple "law of nature" (in the sense of a regularity in the structure of our physical subcomplex) (e.g., the persistence of objects), with a simple mathematical equation (e.g., $1+1=2$), with reference to their relations to the circum- and *I*-complex.

Every time I touch the table with my hand, my perception agrees with my memory-expectation and my expectation with my memory-perception and the "law" of the persistence of my table is an expression of this reciprocal relation which we can view as a connecting of the *I*- and circumcomplex. But this relation is itself *independent* of its accompanying expression: I can experiment with my table and experience anew its presence even *without* verbal thoughts.

The "one plus one" is *not* the expression of a physical regularity but of a linguistic one, and without verbal thoughts the equation cannot be tested on its true experiential content (the experience that every man names "one and one" "two").

It is a pity that in daily life, natural law and mathematical equation are not as precisely separated as the signific theory would demand. Physical and verbal regularities mutually influence each other. Thus the $1+1=2$ and the "natural law" of persistence are closely tied together. If instead of touching a table repeatedly I touch a pair of gloves or shoes, then I express the "persistence" of this pair (in an unphilosophic manner) by the formula that "one and one" always show themselves to be "two." Arithmetic equations thereby acquire a twofold character. That $7+13=20$ can be tested in two ways—by inquiry and by counting— but in one case we test an *equation* in the other a *natural law*.

This twofold character is attached to all applied mathematics. Consider infinite and null aggregates. In other words, wherever a definition demands a *negation of exclusion* no physical correlation is possible. This is due to the fact that the negation of exclusion is distinguished from negation of choice by its *emotional* meaning elements. The infinite has a purely formal significance in mathematics and a purely emotional (or volitional) one in everyday language, and the "law of excluded middle" is not capable of application in physics. The failure to distinguish these two meanings has frequently led to confusion of concepts. Such a confusion lies in the question of the "existence" of the actual infinite.

A frequent confusion of ideas related to this is connected with the concept of true contradiction or self-contradiction. In daily life self-contradiction is merely ambiguity of the expression of the will, and the boundary between "ambiguity" and "uniqueness" is weak. The expression of the will can carry with it more or less satisfaction. In formalized language (logical or mathematical) *no* gradual transition from contradiction to noncontradiction is possible: $a \neq -a$ and *only* $a \neq -a$ is the equation of contradiction when we give the symbols their usual meaning, i.e., when they satisfy the usual rules of calculations.

Now the difficulty arises when the question of *freedom* from contradiction is brought up relative to an *infinite* system. It is clear that this concept has just as little linguistic experiential content as the concept of the actual infinite has *physical* content. The question whether in an essentially infinite system (i.e., a system which cannot be defined by means of a finite expression without using a negation or exclusion) no contradiction can occur "at all" is not an undecidable question—it is *no* question at all. Brouwer's mathematics by freeing itself from the law of excluded middle is much closer to living language than the logistics which has arisen through the formalization of the classic logic.

It is necessary to attempt to say what mathematics is by projecting a formulation of its limits starting from the proper verbal associations and to compare the project with the experiential content related to it. In this sense Mannoury affirms that we can speak only of formalistic or formal-mathematical language when the linguistic reaction of the hearer to the verbal act of the speaker is, as much as possible, independent of the persons and of the accompanying phenomena. It becomes obvious at once that this formulation is applicable only to definite *groups* of persons, and even with this limitation embraces much more than we ordinarily mean by mathematics. A rather large group of Europeans will give the same answer to the question, "What is the first letter of the alphabet?" as they would to the question, "What is 3 + 5" with the same sound. But only an extremely small number of investigators would consider the first question to be a mathematical one. A further limitation must be made which could hardly be found elsewhere than in the *possibilities of the application* of mathematical language.

But the possibilities of application are partially of a general psychological nature and make it necessary therefore to give a description of the concepts of the conceptual forms of mathematics (or intuitionist mathematics), apart from their application. For this purpose it is proposed to count as formalistic mathematics only those forms which satisfy the criterion mentioned above and which can be used as an expression of "mathematical sequences (Folgen)" in Brouwer's sense. Conversely, intuitionist mathematics will be viewed as the totality of these mathematical sequences (viewed as mental chains of association) in so far as they are accompanied by the verbal aids described above. There remain then residual fields which can be called extra-mathematical formalism or extra-mathematical intuition. As examples of the former the letters of the alphabet and the game of chess are cited, and of the latter, morals and spatial imagination are examples.

Since mathematics is relative to linguistic acts and verbal forms its development is also relative to the development of human language and thoughts. It takes its (relatively!) "universal validity" only from this analysis. Everything else which men have tried to ascribe to mathematics—its absoluteness and complete exactness, its generality and autonomy, in a word its truth and eternality—are only superstitions. Mathematics is completely relative!

Mannoury's position may be summarized by following Heyting's remarks.[9] For Mannoury, mathematics must be considered as a phenomenon of life and hence cannot be independent of human beings and their purposes. He then distinguishes between the *forms* in which mathematics *occurs* (formalistic mathematics) and the mathematical thought form (Denkform) (intuitionist mathematics).

From this point of view, it follows that for Mannoury the decisive method to use in considering the foundations of mathematics is the *empirical* and especially the *psychological* method. Thus signific or psycholinguistic investigations must precede axiomatic investigations.

---

[9] Heyting, *Math. Grundl.*, pp. 58–59.

CHAPTER FOURTEEN

THE STRUCTURE OF THE MATHEMATICAL SYSTEM[1]

So far we have been considering the general nature of mathematics, and the definition and extension of the number system. This has led to the discussion of the foundations of mathematics and the relation of mathematics to logic and to the thinker. We have therefore been involved in discussions which may have caused us to lose sight of the structure itself. For this reason we again look at the body of mathematics and analyze its formal structure. We shall start with a description of what we might call the "brickwork" of mathematics.

When we open a text book of mathematics we note that it is composed of sets of symbols. These symbols are of two general types: (1) words, (2) symbols proper. In other words, every mathematics text seems to have two parts. The "word" part is not, properly speaking, part of the text except where words are substituted for technical symbols. The complete formalization of the structure of mathematics is an attempt to eliminate all these words which are not part of the structure proper. Words are used in texts to make more intuitive and comprehensible the significance of the symbols to a beginner. The expert has less and less need of such outside assistance. Even a formalization such as is found in *Principia Mathematica* is composed of these two parts.

This distinction between the two types of symbols can be illustrated in the following paragraph:

"30. *Graphical solution of equations containing a single unknown number.* Many equations can be expressed in the form

$$f(x) = \phi(x) \ldots (1)$$

where $f(x)$ and $\phi(x)$ are functions whose graphs are easy to draw. And if the curves

$$y = f(x), \; y = \phi(x)$$

intersect in a point P whose abscissa is ξ, then ξ is a root of the equation (1)." (From Hardy, *A Course of Pure Mathematics*, 7th Ed., Cambridge, 1938, p. 60.)

[1] Cf. Tarski, *Einführung in die Math. Logik*, Vienna, 1937 (Also translated as *Introd. to Logic*, New York, 1941); Cooley, *Primer of Formal Logic*, New York, 1943.

We note that the paragraph is composed of different symbols. One set has combinations of what are called letters of the alphabet—*a*, *b*, *c*, . . .—forming words of the English language, and another set is composed of letters of the English alphabet, letters of the Greek alphabet, and another form of symbol composed of (,), =, 30, etc.

These symbols are used to represent certain definite ideas and they may be used to represent themselves. In so far as we analyze the properties of these symbols as mere symbols we enter the realm of logical syntax. That is, the study of these symbols in their function as a language leads to the syntactical analysis of the *language* of mathematics.[2] But in so far as we analyze the symbols as representative of certain definite concepts found wherever one speaks of *mathematics*, we are dealing with the *foundations* and *philosophy* of mathematics. As we have seen, it has been maintained that mathematical foundations concern themselves with the properties of symbols as such, and also with the meaning and foundation of specifically mathematical concepts.

The analysis of mathematical symbols which we shall now make will be based on a more operational approach than either of the two extremes mentioned. We classify the symbols of a mathematics from the point of view of the way they are used and shall consider some of the ideas associated with these uses.

Mathematics is constructed from sets of symbols and combinations of these symbols. Some of the combinations are seen to be replaceable by other combinations under definite conditions, others are not. Some combinations and even single symbols are said "to follow" or to be "implied by" others, while some are not.

If we consider single symbols, we find that some of them, e.g., 5, 1, (,); etc., have unique meanings. If they are replaced by other symbols these new symbols can only be alternative ways of saying the same thing. Quite frequently (e.g., replacing 5 by a letter, say *a*) new insight can be gathered into the meaning of a combination of symbols in which the single symbol occurs. But no new entity occurs. Such a symbol is called a *constant*. In terms of *classes* we can say that a constant denotes a class (or a proper subclass of a class) containing one and only one element. There seems to be no special difficulty in the nature of a constant. The problem there is merely that of symbolic reference.

There occur symbols in addition to those indicated (e.g., *x*, *y*, etc.) which do not have unique referents. For example, "*x* is a prime number." Here the *x* may be replaced not merely by 3, but also by 5 and

[2] Cf. Carnap, *Logical Syntax of Language;* also the work of Church, Kleene, and Rosser referred to earlier. The various aspects of symbols from which analyses can be made are discussed in C. W. Morris, "Foundations of the Theory of Signs," and Carnap, "Foundations of Logic and Mathematics," both in *International Encyclopedia of Unified Sciences*, Vol. 1, No. 2 and No. 3, Chicago, 1939. *See also* C. W. Morris, *Signs, Languages, and Behavior*, 1946.

we may still get valid, meaningful propositions. The entire phrase "x is a prime number" we called earlier a propositional function whose *variable* is x. This means that every *function contains* a variable (or variables). Variables enable us to set up general propositions and discover general laws. This is due to the fact that a variable denotes a general term and therefore the proposition in which it occurs becomes a general proposition.

Every combination of symbols which occurs in mathematics will be composed of constants, variables, or combinations of constants and variables.

Many textbooks in mathematics use the definition of a variable which goes back to Newton, i.e., that a variable is a quantity which changes. This definition is clearly absurd, for one of the primary conditions for meaning is the uniqueness of the reference of symbols. If, then, a given symbol is permitted to change its reference it clearly would destroy the conditions for meaningfulness. In a sense then, even a variable must remain constant, i.e., its referent must not be altered.

The element of indeterminacy has also been used as the definition of a variable. "A variable," says Couturat, "is an indeterminate term for which we can substitute any determinate term (belonging to a certain class)."[3] This also is not very adequate since the term *indeterminate* is itself vague. The indeterminacy extends only to a definite point. No variable is so indeterminate that it does not indicate a definite kind of element which can replace it. The structured whole in which the symbol occurs gives a degree of determination to the variable.

It must always be remembered that a symbol gets meaning in a context. No mark in itself either means anything or is a constant or a variable. A variable is not indeterminate, it denotes any one of a number of definite entities. "A symbol is said to be a variable in mathematics, if it is used to denote any one of a certain set of mathematical objects; *which* of these objects it denotes being left completely indeterminate."[4] The student should realize that sometimes a symbol ordinarily used as a variable may occur in a larger context so that it behaves as a constant. For example, x is a variable, as is x^2dx, but $\int_0^1 x^2dx$ as a whole behaves as a constant, and the ability to substitute for x is greatly restricted. Such an x occurring in this fashion is called an *apparent* (or *bound*) variable.

---

[3] Couturat, *Encycl. Phil. Sciences*, Vol. 1, "Logic," p. 148.
[4] Black, *Nature of Math.*, p. 50.

As an illustration of the use of variables, we set up the following.[5]

Suppose we let *a*, *b*, *c*,. . .*k* be a set of constants such that: *a* is happy, *b* is happy, . . .*k* is happy. Now in place of the separate propositions, we can write, "*x is happy*," where *x* may denote any one of *a*, *b*,. . .*k*. Which one *x* denotes will be determined by additional circumstances. Where the conditions are not further specified, it is not merely because *x* is indeterminate but because it is irrelevant for the purpose. The important element in a variable is that substitution of a definite kind is permissible.[6]

The kind of elements that may be substituted for a variable is determined by the context (i.e., the combinations of other symbols) within which the variable occurs. The combination of symbols "*x* is happy" has significance only within a context. Of course, the word "happy" at once puts the expression in its context and we know that we can substitute for the *x*, individuals belonging to the animal kingdom. Any other substitutions would result in a meaningless (=non-permissible) set of symbols. Again, if we open a mathematics text and find "$x^2+5=7$," we know that for the *x* we can substitute *numbers*. Any other substitutions (e.g., cats) would give rise to a meaningless set of symbols. "The values of a variable are in general restricted to a species of objects determined by the sense of the function symbol."[7] We shall call the class of objects giving rise to *meaningful* sets of symbols, the *range* of the variable. In the case of predicative propositions—having only one variable— the range is determined by the predicate.

Where two (or more) variables occur, the range of one variable depends on the meaning selected for the other. For example, if we say "*x* is a lover of *y*," the ranges of *x* and *y* may be identical people. However, the context may make one of the ranges differ—e.g., when we are talking about people who love good food. In mathematics for the most part the ranges of the variables remain numbers, but need not. For example, "two *x*'s determine a *y*." In one context the *x*'s may be points and the *y* a line; in another the *x*'s and *y*'s may be even numbers.

If we consider the class of all possible symbols, the statement of a function at once divides that class into two subclasses: (1) the class of those symbols which when substituted give rise to nonsensical sets of symbols; (2) the class of those symbols which when substituted for the variable give rise to sets of symbols meaningful in the context under

[5] Cf., for a good elementary discussion, Bennett and Baylis, *Formal Logic*, New York, 1939, Chap. 10.

[6] Cf. Carnap, *Logical Syntax of Language*, pp. 190 ff.

[7] Hilbert U. Ackermann, *Grundzüge der theoretischen Logik*, p. 45. *Also* Black, *op. cit.*, p. 64. A great deal of what follows is due to C. J. Keyser, *Mathematical Philosophy*, New York, 1922.

consideration. The subclass under (2) is what we have called the *range* of the variable. All variables, then, are restricted to a *range*. This range may vary quantitatively from no elements up to an infinite number of elements. This means that every function determines a class (its range). If the variable's range is a unit class, we call it a constant. Thus we can treat a constant as a variable whose range contains one and only one element. From this point of view the distinction between real and apparent variables is a nonmathematical distinction and hence unnecessary. Thus the x in "$x+2=5$" is a constant; also the x in $\int_0^1 x^2dx$ behaves as a constant. The addition of $\int_0^1$ as part of the complete symbol defines a unit class.

From what has been said it is apparent that given a variable occurring in a function in a given context, the variable has a definite range. In other words the "variation" of the variable is restricted. No variable can take on meaningful values outside its range. Hence no variable which is completely a variable exists. Only in such functions as define a subclass under (1), which is null, can the variable take as its values any symbol whatsoever.

Now consider the function $x^2-4x=12$, which we shall assume is found in a mathematical context. The subclass of symbols which can be meaningfully substituted for the variable x is obviously the class of numbers. Any element of this class can be put in the place of x. But we notice at once that many elements of the class of numbers when put in the place of x result, after the operations indicated are concerned, in a set of symbols which do not validly go together. For example, let x be replaced by $+2$, then we have $4-8=12$, which is not so in the system of arithmetic. If, when we replace a variable by a symbol, we get a statement asserting a fact that is not possible within the system under consideration, then we shall call that symbol a *falsifier*. All other symbols will be called *verifiers*. Thus every variable occurring within a function will give rise to two subclasses of meaningful substitutions: (1) falsifiers, (2) verifiers.

Any attempt to show by the rules of logical inference or substitution or transformation (1) that a given symbol or set of symbols is a verifier of a given function, (2) that any verifier of a given function is also a verifier of another given function, (3) that the verifiers of a given function are a subclass of the verifiers of another given function, will be called a *proof*. If a combination of symbols is *proven* on the basis of a set of axioms, the combination of symbols is called a *theorem* of the axioms. If a theorem has been derived, the set of symbols constituting the theorem is said to be *true*. Hence (as Jörgenson points out[8]) a "proof" is the derivation of the truth of a set of symbols by showing it to be derivable from other sets of symbols.

A set of symbols may be *derived* from others either directly or indirectly. (For convenience we shall call any set of symbols which (1) is used as axiom or (2) has been derived from other sets of symbols, a proposition.) Propositions may be obtained, then either by direct methods or by indirect methods. The most elementary form of direct derivation is the rule of the syllogism, which we studied earlier. This means that if we can find a proposition "p implies q" (q being the theorem to be derived), and at the same time p can be demonstrated to be a proposition in the system, then q is said to be derived and hence a proposition. For example:

$\frac{a}{b}$ is an even number *implies* $\frac{a}{b}$ is exactly divisible by 2.

$\frac{a}{b}$ is an even number

$\therefore \frac{a}{b}$ is exactly divisible by 2.

The general form is clearly

$$\begin{array}{l} p \rightarrow q \\ \underline{p \quad\quad} \\ \therefore q \end{array}$$

(another form of proof commonly used and related to this has the form

$$\begin{array}{l} p \rightarrow q \\ \underline{q \text{ is false}} \\ \therefore p \text{ is false}) \end{array}$$

A proof will usually consist of a chain of syllogisms of the form given, the final step of which will be the proposition being derived. A simple example may be taken from Hardy (*Course of Pure Mathematics*, 7th ed., p. 132):

To prove: If $\lim\phi(n) = a$ and $\lim\psi(n) = b$ then $\lim\phi(n) . \lim\psi(n) = a . b$

[8] Jörgenson, *Treatise of Formal Logic*, Vol. 3, pp. 103–34. Other Literature: Holder, *Math. Methode*, pp. 22–28; Hardy, "Mathematical Proof," *Mind*, Vol. 38 (1930), p. 345; Dresden, "Mathematical Certainty," *Rev. Meta. Morale*, Vol. 45; E. T. Bell, "On Proofs by Mathematical Induction," *Amer. Math. Monthly*, Vol. 27, p. 413; also "Place of Rigor in Mathematics," *ibid.*, Vol. 41x (1934), p. 599.

Proof: Let $\phi(n)=a+\phi_1(n)$, $\psi(n)=b+\psi_1(n)$
so that $\lim\phi_1(n)=0$ and $\lim\psi_1(n)=0$.

$$\begin{aligned}\text{Then } \phi(n)\psi(n) &= [a+\phi_1(n)][b+\psi_1(n)] \\ &= ab+a\psi_1(n)+b\phi_1(n)+\phi_1(n)\psi_1(n).\end{aligned}$$

$$\therefore |\phi(n)\psi(n)-ab| \leq |a\psi_1(n)|+|b\phi_1(n)|+|\phi_1(n)\psi_1(n)|$$

from which it follows that

$$\lim\{\phi(n)\psi(n)-ab\}=0$$

from which the theorem follows.

Another method of obtaining theorems directly is by substitution of special cases for general variables.[9]

Of indirect methods, we have two important techniques: (1) *reductio ad absurdum*, (2) induction.

The *reductio* (reduction to absurdity) method really proceeds by the rule of syllogism but its essence consists in proving the absurdity of assuming the proposition to be derived to be false. A simple illustration of this method may be taken from elementary plane geometry:

To prove: Two parallel lines can never meet. (Parallel lines are here defined as two lines, such that, when they are cut by a transversal, the angles formed with the given lines on the interior of and on the same side as the given lines add up to 180°.)

Proof: Draw any line intersecting the two parallel lines

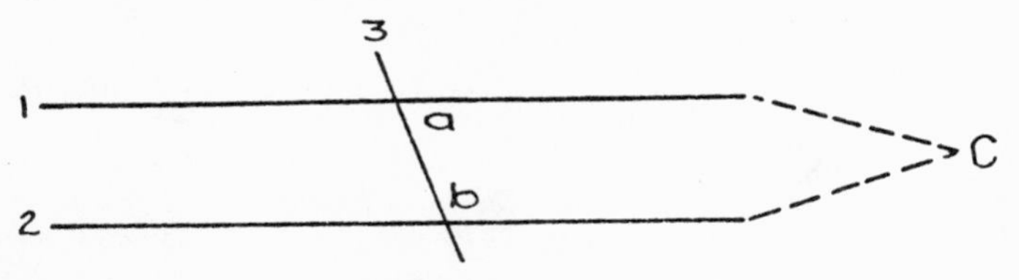

Now assume lines 1 and 2 do meet somewhere at a point C. Then we would have $\angle a+\angle b+\angle c=180°$. But we know $\angle a+\angle b=180°$. Therefore $\angle c=0°$—which is absurd. Therefore the assumption that the lines do meet, leads to an absurdity. Therefore they cannot meet.

[9] The reader should consult some textbook in Logic for the theory of Syllogistic (direct) inference if he is not already acquainted with the theory; e.g., Cohen and Nagle, *Introduction to Logic and Scientific Method*. Cf. Tarski, *op. cit.*, pp. 26–27; also Carnap, "Foundations of Logic and Mathematics," *International Encyclopedia of Unified Sciences*, Vol. 1, No. 3, pp. 12–18.

The method of induction[10] consists of two steps:

1) the theorem is true if when it is true for *any* value whatsoever, it is true for the *successor* of that value.

2) the theorem is true for special cases.

There are other types of methods but these are the most important of them. Mathematical proof seems to be a manipulation of sets of symbols in certain definite ways in order to attain the proposition desired.

New combinations of symbols are frequently introduced into a mathematical proof by the words "definition" or "let. . . ." Thus we have frequently run across such expressions as "let $\phi(n)=a+\phi_1(n)$" or "by definition $\frac{df(x)}{dx}=\lim_{\Delta x\to 0}\frac{f(x+\Delta x)-f(x)}{\Delta x}$." In both cases we introduce a new proposition consisting of a symbol which did not occur previously (at least with the meaning now given it). Where a symbol is introduced in such a way, we say we have a *definition.*[11]

Definitions are introduced for specific purposes. Frequently it is desirable to substitute a new symbol for a group of symbols which has become unwieldy. The purpose here is merely one of ease and convenience. It is the "shorthand" purpose. Such a definition is called a *nominal* definition. In this definition we are really saying that the symbol combination $\frac{df(x)}{dx}$ occurring on the left-hand side of the equality sign will be used for the more complicated symbol combination occurring on the right-hand side. This kind of definition, as any definition, enables us to convert sentences in which the $\frac{df(x)}{dx}$ occurs into sentences in which it does not occur, and also sentences in which the $\lim_{\Delta x\to 0}\frac{f(x+\Delta x)-f(x)}{\Delta x}$ occurs into sentences in which it does not occur. In other words, definitions enable us to remove from our language certain sets of symbols (or words). Thus in geometry we can remove the symbol "line" (or "point") by defining "line" as the shortest distance between two points (or "point" as the intersection of two lines). This is the most important function of a definition.

---

[10] Cf. G. E. Raynor, "Mathematical Induction," *Amer. Math. Monthly*, Vol. 33, p. 376; also Bell, "On proofs by mathematical induction," *ibid.*, Vol. 27, p. 413; Russell, *Introd. Math. Phil.*, Chap. 31; Dubislav, *Phil. d. Math.*, pp. 28–31.

[11] Cf. W. Dubislav, *Die Definition*, Leipzig, 1931, for a complete discussion of the theory of Definition. Also, Weyl, *Phil. d. Mathematik*, *op. cit.*, pp. 8–9; Tarski, *Einfuhrung*, pp. 19–22; Stebbing, *Modern Introd. to Logic*, p. 421.

A second type of definition which occurs in mathematics may be called *operational definition.* This defines the symbol by means of the ways in which it combines with other elements. For example, we may define the number 0 by the two conditions

$$n+0=n$$
$$n\times 0=0$$

This kind of definition occurs also in the definition of elemental concepts in postulate sets. Thus "natural number" may be defined by means of some postulate set for natural numbers such as that given in the chapter on the postulational definition of number (pp. 58–62). This type of definition is sometimes called "systemic" definition.[12]

When we define a new symbol, we must not set up a definition which will lead to contradictions. To use an illustration of Peano's:[13] Suppose we attempt to introduce a new symbol "?" into arithmetic by the definition

$$\frac{a}{b}?\frac{c}{d}=\frac{a+c}{b+d}$$

We cannot accept this because if $a=2$, $b=5$, $c=7$, $d=11$ we get

$$\frac{2}{5}?\frac{7}{11}=\frac{2+7}{5+11}=\frac{9}{16}$$

and if we replace $\frac{2}{5}$ by its equal $\frac{4}{10}$ we get

$$\frac{4}{10}+\frac{7}{11}=\frac{4+7}{10+11}=\frac{11}{21}\neq\frac{9}{16}$$

The introduction of a new symbol by means of the attribution of a definite property to an operation (e.g., $a+b=b+a$) must not lead to contradictions.

Another illustration of this type of definition is the definition of a *matrix.* In order to define a matrix the following set of definitions (really constituting a single definition) are laid down:[14]

1. A system of $m.n$ quantities arranged in a rectangular array of m rows and n columns is called a matrix.

---

[12] Cf. L. O. Kattsoff, "Undefined Concepts in Postulate Sets," *Philos. Review,* Vol. 47 (May, 1938), pp. 293 ff.

[13] Peano, "Les definitions mathématiques," *Bibliothèque du Congrès Int. de Phil.,* 3, 1901.

[14] Taken from M. Bôcher, *Introd. to Higher Algebra,* New York, 1929, pp. 20 and 61–63.

2. Two matrices are said to be equal when and only when they have the same number of rows and of columns, and every element of one is equal to the corresponding element of the other.

3. By the sum of two matrices of m rows and n columns each we understand a matrix of m rows and n columns, each of whose elements is the sum of corresponding elements of the given matrices, etc.

A type of definition which is of frequent occurrence is illustrated by the following:

A *bilinear form* is a polynomial in the $2n$ variables $(x_1 \ldots x_n)(y_1 \ldots y_n)$ where each term of the polynomial is of the first degree in the x's and also of the first degree in the y's..

A *straight line* is the shortest distance between two points.

A *singular matrix* is a square matrix whose determinant is zero.

In these definitions we define special subclasses of elements of an already defined or known class. The definitions consist of naming the genus, and adding a specific property. Thus a singular matrix is put into its genus *square matrix* and the additional property of having its determinant zero is added. Notice that we already are presumed to know the meaning of *square matrix*, *determinant*, *matrix*, and *zero*. This is the classic form of definition and may be called *formal definition* (or definition by genus and difference). It was thought by Aristotle that all definitions must be in this form.

The first part of the systemic definition of "matrix" creates a new entity by combining a group of symbols already at hand. This type of definition is called a *combinatoric* definition. Its purpose is to give us a symbol for a specific combination of old elements. It is in a sense practically the same as the nominal definition. It differs from the nominal definition in that the symbol represents a new entity which has properties not present in its parts. Another illustration of a combinatoric definition may be found in the definition of *cross-ratio*.[15]

"If L_1 L_2 L_3 L_4 are four distinct concurrent lines of which L_1 and L_2 are finite, the cross ratio $(L_1\ L_2\ L_3\ L_4)$ is defined as the quotient

$$(L_1L_2L_3L_4) = \frac{\text{Ratio in which } L_3 \text{ divides } L_1,\ L_2}{\text{Ratio in which } L_4 \text{ divides } L_1,\ L_2}\text{''}$$

It is evident that in a sense we are merely introducing a symbol to stand for a more complicated set of symbols—but we are doing more. A new entity is constructed.

[15] W. C. Graustein, *Introd. to Higher Geometry*, New York, 1933, p. 72.

Sometimes, however, the new entity is not merely constructed by a combination of symbols already at hand but is an entirely new entity. Usually the entity created is an ideal one—a limiting case. A definition which thus *creates* a new entity is called a *creative* (schöpferische) *definition*. Such a definition is usually introduced to make possible generalizations or to make possible proofs of theorems which might otherwise be extremely complex. An example of such a definition is that of points at infinity.

"Corresponding to each set of lines consisting of all the lines with a given direction, there is created an ideal point, or point at infinity."[16]

With this point at infinity it is possible to assert that any two lines, even parallel lines, intersect. This point is in a sense actually created. However, it is possible to consider creative definitions as rules of operation stating how we are to treat this new entity. Parallel lines might be treated *as if* they intersected at infinity. But we can give properties of this point and hence it would seem to have some kind of existence.

Another example of the same type of definition occurs in the theory of functions. Frequently a function is created to prove or disprove a given theorem. This definition does not tell us what the function is but merely establishes a correspondence. This is really a partial construction of the function. For example, to show that some functions exist in which either the right-hand or left-hand limit or both fail to exist at a given value of the variable, the following function is defined

$$f(x) = \sin\frac{1}{x}, \text{ when } x \gtrless 0$$

$$f(x) = 0 \text{ when } x = 0$$

At the origin a discontinuity of the type desired exists.[17] The form of the $f(x)$ is not given.

An aggregate, being always a creation, i.e., a combining of elements to form a new entity, is thus defined by means of creative definitions. It is clear that a great deal of the advance in mathematics is due to creative definitions. Certainly, whenever we define new entities by means of abstractions, we are using this type of definition.[18] (We consider abstractive definitions to be special cases of constructive definitions and shall not discuss them further.)

We have distinguished five main types of definitions: (1) nominal, (2) systemic, (3) formal, (4) combinatoric, (5) creative. All definitions as we have seen, are combinations of symbols. In each case a new

---

[16] Graustein, *op. cit.*, pp. 21 ff.

[17] Townsend, *Fun. of Real Var.*, p. 99.

[18] Cf. Schoenflies, "Uber die Stellung der Definition in der Axiomatik," *Jahres. Deut. Math. Ver.*, Vol. 20, pp. 227–55.

symbol is introduced and related to a group of symbols which have already occurred in the system. Each of these symbols which serve to define the new symbol either occurred before in a definition or was present in the initial combinations of symbols from which our system had its origin. All symbols not defined by means of nominal, formal, combinatoric, or creative definitions are called *undefined symbols* and represent undefined concepts.

Since these symbols, as is obvious, are defined (in the sense that their properties are stated) systemically, it is perhaps better to call them *initial symbols*.[19] It is a combination of these *initial symbols* with symbols defined or derived from these initial symbols that enables us to construct the entire system. We have certain combinations of initial symbols, one or more of which is a relation. From these combinations which are laid down, we start our system by calling them axioms or postulates. These are called also undemonstrated propositions. To avoid ambiguity it would be advisable to call these *initial* propositions rather than undemonstrated.' It is obvious that these propositions justify the system and from that point of view are really demonstrated systemically.

The initial symbols are found to be of three kinds as they occur in initial propositions: (1) term symbols, (2) operation symbols, (3) relation symbols. Combinations of these in accordance with proper syntactical rules give rise to propositions, either initial or proven.

Term symbols are used as are nouns in ordinary language. Examples of term symbols are 1, 2,. . . symbol for a matrix, symbol for a derivative, line, point, etc. Term symbols may be composed of one or more symbols. Thus $f(x)$ is a term symbol composed of the relation symbol f, the term symbol x, and the operation symbol ( ). It is evident that a term symbol may represent either a constant or a variable. Term symbols indicate the presence of concepts in logic and ordinary language. When we seek an interpretation of a system of mathematics, term symbols are usually interpreted by the use of nouns.

The analysis and establishing of the term symbols of mathematics constitute an important branch of the philosophy of mathematics. There are three fields of investiagtion which develop:[20] (1) The relation of the term symbol to the action state and environment of the man who sets them up. This is called *pragmatics* and is the field of investigation of Mannoury and Pasch as studied in earlier chapters. (2) The

---

[19] Cf. the discussion of this problem in Kattsoff, *op. cit.*, pp. 293–300.

[20] For analysis of theory of signs, *see* Morris, "Foundations of the Theory of Signs," *International Encyclopedia of Unified Sciences*, Vol. 1, No. 2; also Carnap, "Foundations of Logic and Mathematics," *op. cit.*, Vol. 1, No. 3; and Carnap, *Introduction to Semantics*, Harvard Univ. Press, 1943.

relation of the term symbol to that which it designates. This is called *semantics* and is the field of investigation of Frege, Russell, Brouwer, and the earlier more metaphysical approaches to the definition of number. (3) The relation of the term symbol to other symbols in the system, abstracted from any extra-linguistic reference. This is called *syntax*—a term which describes clearly the field of investigation. This is the field of investigation of Hilbert, Chwistek, Church, and Carnap. (These three fields of investigation can be distinguished in the study of all types of symbols, not merely term symbols—although pragmatics, especially, and semantics seem to be more relevant to term symbols.)

Any symbol or combination of symbols may become a term symbol, especially if we are concerned with its syntax. Term symbols form the elements which give significance to the operation and relation symbols. These latter never occur alone (except in syntax when they are treated as term symbols) while term symbols may occur alone or in combination with either or both of these. (For example, $2+2>2$. The term symbol 2 occurs here together with the operation symbol $+$ and the relation symbol $>$.)

Operation symbols occur in conjunction with two or more term symbols. They denote the method of forming new term symbols from those already at hand by combining the given symbols in definite ways. The most common operation symbols in mathematics are the symbols for multiplication, addition, subtraction, division, and exponentiation. Operation symbols never occur alone if they are to have meaning. The usual symbols in mathematics for the first four of these are $\times$, $+$, $-$, $\div$. The application of these operations to two or more term symbols will give us another term symbol. For example, $2+2=4$; $8\div 4=2$; $f(x)+\phi(x)=\psi(x)$, etc.

Other operations which occur are integration, differentiation, substitution, transformation, inversion, etc. Many of these are indicated by the use of unique symbols, others are not. When we have a new term indicated by an operation on two terms for which we have no term symbol, a term symbol may be introduced by means of a nominal or combinatoric definition. The extension of the number system was seen to result because of this fact. For example, the operation of subtraction on 3 and 5 (i.e., $3-5$) necessitated the introduction of a new term symbol -2, and thus arose the extension of the number system to include negative numbers. (Note the use of the operation symbol as part of the term symbol -2. In this context "$-$" is meaningless by itself.)

The last type of initial symbol we shall discuss is the relation symbol. Involved in the relation symbol is the entire theory of relations of logic. A relation also occurs in conjunction with term symbols. But unlike operation symbols, it does not create new term symbols.

It merely establishes a comparison between the term symbols. Illustrations of relation symbols are equality ($=$), greater than ($>$) and its converse less than ($<$), between, etc.

A special and very important type of relation is involved in the *function* concept. A relation is said to be a function if it denotes a correspondence between the members of two or more *classes*. Thus the relation $>$ may become a function if we use a variable in one or more of the places in which term symbols occur in this relation. For example, $x>2$ (denoted $>(x, 2)$) is a function whose values are truth-values. Other illustrations are $x^2=y$, $\cos y=x$, etc. The general form of a function is $f(x)=y$ or $f(x,y)=z$. This can be generalized. In every case, however, the f establishes a correspondence between two classes of term symbols. A function may be viewed as a generalized relation. When constants are substituted for the variables in a function the result is a relation.

Every relation establishes a connection between two or more terms. When the connection is between the terms of two or more classes (which may be different or identical), it is a function. This means that every relation defines a given set of terms, namely, the set of terms so related. Thus $3>2$ is a relation establishing the set of terms (2, 3). "Point A is between points B and C" establishes the set of terms (A, B, C). In the case of a function, e.g., $x>y$, where x and y take an element of the classes of numbers (1, 2, 3) and (2, 4), the function establishes the set of relations $2>1$, $4>1$, $4>2$, $4>3$ and therefore the class of couples $\{(2,1), (4,1), (4,2), (4,3)\}$. The terms which are made to correspond are united to form a new class which is defined by the relation (or function).

In the case of a function, the new class consists of the verifiers of the variables. A function in which the relation holds between two terms will define a class of couples. In general, a function in which the relation holds between n-terms will define a class of n-tuples $\{(a_1 \ldots a_n), (a^1_1 \ldots a^1_n) \ldots (a^m_1 \ldots a^m_n)\}$. For example, the function $x^2+y^2=z^2$ will define a class of triplets. Subclasses of this class of triplets can be obtained by holding one of the variables constant. A subclass of the class of triplets defining $x^2+y^2=z^2$ can be obtained by letting $x=2$ and then define the class of couples of $4+y^2=z^2$.

It is clear that solutions of functional equations of n-variables are really the discovery of the class of n-tuples. Thus in the case of Fermat's theorem $x^n+y^n=z^n$ for $n>2$, the class of n-tuples has not yet been discovered. Perhaps the class defined by the equation is a null class.

Two types of functions occur in mathematics. One is the type we have been discussing, which is actually a special case of propositional function. The other type is the set of postulates as a unit. Various interpretations can be given for a set of postulates, hence a set of postulates can be viewed as a function defining a class of classes of elements. Such a function is called a *doctrinal function* (Keyser, *Math. Philosophy*). Doctrinal functions are important since their domain of possible interpretations may include empirical data. The application of mathematics to reality can be understood with the aid of this concept.

A relation symbol of tremendous importance both in logic and mathematics is the symbol "=" (read "is equal to" or "is identical with"). A great many proofs in mathematics are attempts to reduce an original assumption to the form $a=a$. This symbol varies in meaning with the context in which it occurs, although the general concept of some kind of replaceability of either side of the equality by the other remains.

In logic the equality (or identity) concept has the following meanings:

Two individuals are said to be identical if every property that one has is possessed by the other. Or more precisely, $x=y$ if every propositional function satisfied by x is also satisfied by y and conversely.

Two classes are said to be equal ($M=N$) if each class is a subclass of the other; i.e., if all the individuals of one are equal to all individuals of the other.

Two propositions are said to be equal ($p=q$) if each implies the other.

In all cases, if equality has been demonstrated either element of the equality may replace the other wherever it occurs. In all of these definitions of *equality* in logic, the central idea is that the two elements involved have the same properties from a given point of view.

In mathematics, again, the central idea is that of substitution. In arithmetic two numbers are said to be equal if either can replace the other wherever it occurs. This means, since we are dealing with magnitudes, that they have the same magnitude. The equality of fractions, complex and irrational, is reducible to equality of natural numbers and hence introduces nothing new.

In algebra, two rational integral functions in x are said to be identically equal if they are equal for all values of x.

In geometry, two angles—or figures or lines—are said to be equal if they have the same magnitude and the same form. This means that we can place one figure upon the other.

With these types of symbols we can, by proper combination, build up the structure of mathematics.[21] It is necessary, of course, to have

[21] An extended discussion and enumeration of mathematical symbols is found in André, *Des Notations Mathematiques*, Paris, 1909.

rules of combination, some of which are expressed by means of the symbols themselves (by writing down an exemplar of the permissible types of combination); others are noted in a language extra-mathematical (metamathematics). Combinations of these symbols, which are arranged according to the syntax of the system, will form *sentences* of the system.

A sentence in which a relation symbol occurs will be called a *proposition*.[22] Thus $2+3, f(x)-5$, etc., are all sentences, while $2=1+1$, $f(x)<\phi(x)$, etc., are that type of sentence in which a relation symbol occurs and is called a proposition. Thus the term sentence includes in this usage the term proposition. The initial sentences of the system are systematically demonstrated. The truth values they have are therefore determined by the system as a whole.

These initial sentences will form the set of postulates of the system and must satisfy the various properties of a set of postulates. The system of symbols comprising the initial sentences and all sentences derived from them constitute the postulational system.

In such an analysis as we have made in this chapter the question of foundations would have no place at which to appear. We make a purely abstract description of the symbols we find in a mathematics text. However, such a description tells us very little of the nature of mathematics, of the origin of mathematical concepts, of the applicability of mathematics, etc. To describe in abstraction the symbols of mathematics is to make of the system a sort of Platonic idea. This does not explain the empirical origin of so much of mathematics. Mathematics does not arise *in vacuo*. It is a cultural phenomenon. A purely syntactical approach to mathematics is refuted by the role of mathematics in society. This, however, in no way denies the extreme importance of this type of investigation of mathematics.

Before considering more closely the relation of mathematics and reality, we turn our attention to the postulate system which has become the model for investigation.

---

[22] This use varies from the usual one. For a more detailed analysis *see* Carnap's works cited above.

CHAPTER FIFTEEN

POSTULATIONAL METHODS[1]

From Aristotle to Bacon and Descartes, the science of mathematics was looked to as the model for all the sciences to follow. Its methods are reputed to be applicable to all knowledge and to be the source of truth and discovery. Since of all branches of mathematics, geometry was the best developed, its methods were taken to be the method of all mathematics. Schoenflies says (*Math. Ann.* 83) that Hilbert's foundation of geometry (a purely axiomatic one) must be the model for all analogous investigations. The form in which geometry appeared was what is now called *postulational*. Hence we have such an attempt as Spinoza's to use this method in the field of ethics. That ethics does need a new method and has needed it since Socrates, is an indisputable fact. It is to Spinoza's credit that he attempted to use a new method, but there was no attempt to analyze the method of postulates to see whether or not it was applicable to such systems. For the most part, even today, the advocates of a postulational technique assume without question the validity of this method.

The reason for this faith in a postulational method is the fact that in such a method one finds exactitude of definition and rigor of deduction at their best. But it must be remembered that to use a set of axioms even in symbolic form does not necessarily imply the introduction of *metric properties*. If it did, there would be great importance to the objection that such a method cannot apply, say, to sociology. As will be evident later, all that the axiomatic method does imply is that we set down in precise form the definitions we are using and the principles of deduction. In this way as little room as possible is left to objections based on misunderstanding of words or other ambiguities. The difficulty involved in precise definition cannot be taken as an excuse for vague terminologies, and is therefore not an argument against the use of an axiomatic. No one knows better than the mathematician how difficult it is to get precise and exact definitions. Yet the mathematician has advanced his science only by dint of continual redefining.

---

[1] The literature on the problems related to Postulational Techniques is now great. A complete bibliography is found in *Journal of Symbolic Logic*, to which the reader is referred. Parts of what follows were published in Kattsoff, "Postulational Methods II," *Phil. of Science*, Vol. 3, No. 1 (January, 1936), pp. 67 ff.

When mathematics is defined in terms of its method it is this postulational technique which is kept in mind. (Reference to the definitions of mathematics given in an early chapter will clearly illustrate this.) The postulational method is the form of all deductive theory and hence seems to lay down the conditions necessary for scientific formulation of a sphere of knowledge. If mathematics is to be considered a science along with physics, then it would appear probable that the common element is the deductive (postulational) form.

In 1899 Hilbert had completed the axiomatisation of Euclidean Geometry and thereby established that science on a firm foundation. The idea that a science founded on axioms is certain and indisputable is not a new one, as has been pointed out, but goes far back. It is quite evident that Euclid's attempt to systematize geometry—which resulted in his Elements—was due to a feeling that there was an absolute science. This feeling was further evident in the efforts of Bolyai, Saccheri, and Lobatchevsky to prove the parallel postulate, since it seemed to be imperative that a certain science should be founded on as few axioms as possible, and these axioms should be as self-evident as possible. In Bacon also we find this emphasis on an axiomatic science. But the real impetus to axiomatic methods was due to Hilbert. In 1900, Hilbert insisted that despite the pedagogical value of genetic methods, preference was to be given to axiomatic methods because they assure a final validity and logical certainty. Again and again Hilbert insists that the only way to rid mathematics of its sore spots and achieve a safe foundation is to use the axiomatic method. It seems natural that Hilbert should rely so strongly on this method, since he has seen it give so safe a structure to geometry.

The use of this method has been extended into the fields of physics, biology, psychology, etc. Probably the importance of the technique lies in its systemic character.[2] It enables one to construct a system out of atomic elements. For this reason it is in line with the methodological consideration in many fields. We have noted earlier that the so-called "undefined concepts" and "undemonstrated propositions" of a system are really defined by the system. No infinite regress making knowledge impossible really exists. Only if we consider a system atomistically, i.e., as built up from atomic elements, does this problem arise. Every definition and every concept introduced into a system is done so in the light of the system we are developing.

The trend of thought in psychology today illustrates the point. Much of contemporary study in psychology emphasizes the need for

[2] For example, S. C. Dodd, *Dimensions of Society;* C. L. Hull, *Principles of Behavior,* as well as *Mathematico—Deductive Theory of Rote Learning*; J. H. Woodger, *Axiomatic Method in Biology*; J. von Neumann, and O. Morgenstein, *The Theory of Games and Economic Behavior*.

a more systemic approach to the psychological study of man. Lewin's topological psychology emphasizes the region, Kretchmer's characterology insists upon the whole individual. Many other examples could be given.[3] The point is that a postulational system considers its subject matter as an interrelated structure. Hence the utility of this method for other fields of learning.

To return to our analysis. If the deductive method is considered to be the essential characteristic of mathematics, then the postulational form is the essential form in which mathematics appears. It will then be possible to mathematicize any field of investigation.

THE ELEMENTS OF POSTULATE SETS

In order to carry out our analysis, it is advisable to have before us a typical set. We take the set of five postulates for Boolean Algebras given by H. M. Sheffer in *Transactions American Mathematical Society*, Vol. 14, pp. 482 ff.

We assume

I. A class K

II. A binary K-rule of combination

III. The following properties of K and "/":

1. There are at least two distinct K elements.
2. Whenever a and b are K-elements, a/b is a K element.
3. Whenever a and the indicated combinations of a are K-elements $(a')'=a$.
4. Whenever a, b, and the indicated combinations of a and b are K-elements

   $$a/(b/b')=a'$$
5. Whenever a, b, c and the indicated combinations of a, b, c are K-elements

   $$(a/(b/c))' = (b'/a)/(c'/a)$$

Definition

$$a'=a/a$$

CLASSIFICATION OF POSTULATES 1–5

Postulate 1 is an existence postulate. Postulate 2 demands that the K-rule of combination shall be K-closed. Postulate 3 which demands that a and $(a')'$ shall always be names for the same K-element—that the names a and $(a')'$ shall always be equivalent—is an equivalence postulate; so are 4 and 5.

---

[3] The clearest example is C. L. Hull, *Principles of Behavior*, 1944.

CONSISTENCY OF POSTULATES 1–5

With the following interpretation of K and $/$, Postulates 1–5 are satisfied: K has only two distinct elements m and n; $m/n=n, m/n=n/m=n/n=m$.

INDEPENDENCE OF POSTULATES 1–5

We shall give only the proof of the independence of Postulate 5, since it involves the matrix method of consistency. The matrix method of demonstrating independence consists in finding an interpretation of the axiom system which satisfies all postulates but the one in question. For example, if an interpretation is found which satisfies all the postulates but the fifth, then the fifth postulate is independent, i.e., not implied by the other four. Such an interpretation must be found for each postulate in order to demonstrate that each postulate is independent of the others.

The logic behind this is as follows: Since "p implies q" is defined as meaning that whatever satisfies p must also satisfy q, if it is possible to find an interpretation of p that does not satisfy q, it cannot be said that "p implies q"—i.e., q is independent of p. It must be kept in mind that under these conditions, it is not meant that if p does not imply q, then p implies the falsity of q. Independence means that neither q nor its falsity is demonstrable—since if the falsity of q were demonstrable from the remaining postulates, the set of postulates including q would be inconsistent, both q and its falsity then being demonstrable.

The method is illustrated by the matrix for Postulate 5. The interpretation given below in the form of a matrix will be seen to satisfy Postulates 1–4 but not Postulate 5.

In 5, K has only three distinct elements—1, m, n; "/" is defined by the following table:

/	1	m	n
1	1	m	n
m	n	n	1
n	m	1	m

To show that Postulate 5 is not satisfied, we proceed as follows:

Let $a=m, b=n, c=1$

then $b/c=n/1$, which, from the matrix gives m.

Then $a/b/c=m/m=n$

$n'=n/n=m$

The left-hand side is therefore m.

On the right-hand side

$$(b'/a)=n'/m=m/m=n$$
$$(c'/a)=1'/m=1/m=m$$
$$n/m=1$$

The right-hand side is therefore 1.

DEDUCTIONS FROM POSTULATES 1–5

We shall reproduce only one of the proofs given.

A. Whenever a and b are K-elements, $a/b=b/a$

Proof: $a/b=[(a/b)']'$ by 3
$=[(a/\{b'\}')']'$ by 3
$=[(\{b'\}'/a)']'$ by 5, (let $b=b'$, $c=b'$)
$=[(b/a)']'$ by 3
$=b/a$ by 3

We have selected this set, rather than a more mathematical one, because in this form it is easier to note the various elements involved. The axiom sets for the various branches of mathematics (e.g., *Geometry*, by Hilbert, *Analysis Situs*, by R. L. Moore, *Projective Geometry*, by Veblen, *Arithmetic*, by Peano, etc.) are all of the same general form.

The elements of which the above construction is composed are:

a. A set of propositions;

b. A set of assumptions;

c. A set of definitions;

d. An unexpressed method of drawing conclusions;

e. Certain properties as "independence" and "consistency" which are proven of a subgroup of the set of propositions;

f. A set of objects satisfying the construction—which is used to demonstrate consistency.

DEFINITIONS

1. *The entire group of elements enunciated is called the axiom system.*

2. *The axiomatic method is the construction of such an axiom system.*

Weyl believes that the axiomatic method consists in collecting the fundamental concepts and relations from which all concepts and theorems of a science can be obtained by definition or by deduction.

Such a statement, however, leaves open the very important questions—What are the fundamental concepts and relations? How are these determined? Russell thought that implication, negation, and disjunction were the fundamental relations, but Sheffer and Nicod derive these from the stroke function, which is more fundamental in the sense that implication, negation, and disjunction can all be defined in terms of it.

3. *The axioms are that subgroup of all propositions from which the remaining propositions are deduced in various ways.*

It is clear that this problem is of a different nature from that of determining the *fundamental* concepts.

4. *By a proposition we mean any ordered collection of symbols having a definite truth value.*

We previously defined a proposition as a collection of symbols, at least one of which was a relational symbol. Since this would mean the assertion that the relation either holds or does not, the acceptance of one of these alternatives would mean that we have given or have found a truth value for the collection of symbols.

5. *By an assumption within an axiom system, we mean a proposition to which we assign a truth value, determined by the system we are defining.*

Therefore, it follows that definitions are assumptions. It also follows as a consequence of our definition that assumptions (or axioms) are propositions and not propositional functions, from the point of view of the system itself. A confusion has arisen in the minds of many due to their confounding an axiom system with the application of such a system. When an axiom system is applied to a definite set of objects, the propositions are either satisfied or not. In other words, in application of axiom systems, the axioms are capable of taking on a truth value which depends upon the objects substituted for the apparent variables. Therefore, the axioms were considered to be functional. The set of assumptions whose truth values are taken at will are called the "axioms" or "postulates" of the set. These constitute a subgroup of all the propositions; the remaining propositions are called "theorems" since they are derived from the axioms. Their truth values are contingent upon the truth values assigned to the axioms and definitions. The entire set is spoken of as "axiom (or postulate) set."

6. *By a theorem, we mean one of that subgroup of all propositions of the system which can be deduced from the subgroup of axioms.*

It follows from our definitions that the axioms together with the theorems make up the whole axiom system.

THE NATURE AND PURPOSE OF AXIOMS

We have previously defined the axioms of a set of propositions to be that subgroup from which the remaining propositions can be deduced. It now becomes necessary to discuss the selection of these axioms and their meaning.

It has probably been noticed that the terms "axioms" and "postulate" have been used interchangeably. This is due to the fact that in dealing with abstract axiom systems, there can be no such distinction as Euclid tried to draw. His distinction is due to the fact that he is dealing with a certain type of object, and he feels it essential to distinguish between propositions accepted as true which are about geometric constructions (as, e.g., the parallel axiom) and those true but not about such entities (as, e.g., if equals are added to equals the results are equal). Analogous to such a distinction might be one between propositions dealing with logical facts (as the syllogism) and those dealing with the entities assumed in the system (as, e.g., $a+b=b+a$). The former might be called postulates, the latter axioms. However, even such a distinction disappears when we come to derive the system of logic.

If a distinction is desired we might use that between a tautology (i.e., a proposition of the general form "p or not-p") and a proposition not a tautology. But this is unnecessary. The axioms within a given system are taken to be always true no matter what objects *within that system* are substituted for the variables. In dealing with general axiom systems, as we do here, one word is sufficient since the two signify the same objects. This is an example of how, starting with two words for apparently different objects, it has come about by generalization that we drop the distinction. We shall use the words interchangeably, meaning by axiom set (or postulate set) those propositions of a system from which the remaining ones are derived.

Suppose now we wish to set up an axiom set. On what basis are we to found our selection of those propositions we are to use as axioms? And what criteria are to determine our choice?

For Descartes and the Rationalist school (as also for Euclid), axioms are self-evident truths incapable of demonstration. The question of self-evidence has been abundantly treated, and it has been repeatedly asserted that any proposition can be taken as an axiom, and hence no axiom is self-evident. But an important element is forgotten. An axiom may be self-evident for a given system of meanings. As Husserl emphasizes, once you get at the "essence" of a proposition it may be "seen" to be necessary. Naturally, to one who has not grasped the "essence" the proposition will not be self-evident.

Whether or not axioms are self-evident, there appear to be four types of approach to their selection: (1) empirical, (2) rational, (3) empirico-rational, (4) utilitarian. Although history shows that some of these methods have been advocated to the exclusion of others, no one of these methods alone is adequate.

The empirical method has its strongest support in the natural sciences. Here an axiom is determined by the consideration of a great many instances. In time it appears probable that a certain law explains them all. This law is then selected as the axiom from which to start. The discovery of new facts contradicting the results deduced from this axiom would necessitate a search for new axioms. There is, therefore, nothing absolute about such axioms since they depend upon the cases investigated. The axioms depend upon whether or not we *see* them in nature and are not cogent to mathematics at all.

Here we ask—what of purely abstract sets? May it not be that in such sets empirical factors are of little importance? Suppose for example, I set up the following set:

1) Let there be at least three objects a, b, c, and a class K of which they are elements $(a \neq b \neq c)$

2) $a+b=b.c$

3) $a>b=a-c$

4) $(a+b)c=(a-c)+b$

Have I not been entirely free from empirical data in the selection of my axioms? The choice of axioms is then entirely arbitrary.

This is the ultrarationalistic position regarding the selection of axioms. Axioms are merely arbitrary fiats of the mind. There are no essential reasons why one should select those axioms rather than any others.[4] It is all a matter of the language one wishes to use—hence Carnap's Principle of Tolerance. There is no reason to select Euclid's parallel axiom rather than any other axiom concerning parallel lines—except consistency. However, the very condition of consistency shows immediately that axioms are not "taken from the air," so to speak, but their selection is subject to certain conditions. It would be impossible, too, for any individual to set up axioms who had not some previous experience to suggest these to him.

When one considers the purpose of axioms in the natural sciences, the empirical criterion of the selection of axioms leads directly to what may be called the utilitarian or pragmatic criterion (Poincaré). This, briefly, states that the choice of one's axioms depends on their utility. In natural science, utility is in part defined in terms of its powers of explanation and prediction of events. There appears a new term,

---

[4] "Den Axiomen kommt also keine logische Notwendigheit zu" says Hölder in *Die mathematische Methode*, Berlin, 1924, p. 358.

hypothesis, which leads to a different point of view. All empirical or pragmatic criteria for the selection of hypotheses are concerned with *applying* axiom systems. It becomes necessary as a result to distinguish clearly between an abstract and an applied system.

In abstract systems, no reference is directly made to empirical objects. As a consequence, empirical considerations alone are not sufficient for the selection of an axiom set. On the other hand, when axiom systems are to be applied to empirical objects, the set cannot be totally independent of empirical considerations. In abstract sets, as was pointed out previously, we use only the word "axiom"—the necessity for "hypothesis" shows at once that we are no longer dealing with such sets.

Actually axioms are suggested to us or controlled by empirical considerations, but are not *determined* by such considerations. As Dubislav points out,[5] the position that axioms are merely statements of empirical results (Mill) does not take into consideration the process of abstraction. Our experiences are searched for *types* of facts. From among these we make our selection with a certain degree of freedom. After such a selection has been made, we test the axioms for the *properties* of consistency and independence, a desirable property in the case of independence, and a necessary one in the case of consistency. Evidently if we had never come across relations which are transitive, we would probably never think of postulating an axiom as "$aRb \, . \, bRc$ implies aRc" unless it were done as a contradictory (or contrary) of another relation as in the case of the parallel axiom.

Another manner in which axioms are selected, which is not directly empirical, but is based upon our experience, is by the generalization process. In this fashion axiom sets for projective geometry, or for analysis situs, have frequently been constructed. If a name is desired for this method, we might use empirico-rational, or rational-empirical, or (as suggested by E. A. Singer) critical.[6]

Suppose we wished to establish a set of axioms for the relation "betweenness." We would examine various cases where this relation held and consider further what we meant by "betweenness" and write down those properties we thought to be essential. Of course, we need not accept any given property as an axiom. We might take the converse, or the contradictory. Here rational considerations begin to enter. It is a mistake to say that the selection of axioms is *purely arbitrary* in dealing with abstract sets. The insistence of the various schools on the different criteria discussed above turns out to be due to the fact that they were considering only separate aspects of axiom systems.

[5] W. Dubislav, "Uber den sogenannten Gegenstand der Mathematik," *Erkenntnis*, Vol. 1 (1930), p. 37.

[6] Langer treats this process very lucidly in *Introduction to Symbolic Logic*, New York, 1938. The student should read this book carefully.

Having selected a set of propositions to use as axioms, we must apply certain criteria to them. The first criterion that comes to mind is a minor one—that of economy in the number of assumptions. We are to select, all other things being equal, the least number of propositions possible from which all others can be derived. Clearly we might (if we had before us the entire system) select all the propositions as axioms, and then have nothing left to prove. This does not take account of our demand. If we have the smallest possible number of axioms, we need not consider the other propositions since they can all be derived from these. If we have a set of m propositions and can deduce from r of them the other $m-r$ propositions, and also from s of them the other $m-s$ propositions, we shall select (other considerations being equal) as our axiom set either the r or the s propositions, according to whether $r<s$ or $s<r$. Clearly if $s=r$ we must consider other criteria. If no other are available we select at random either set s or set r.

This criterion for the selection of our axioms is a method applicable only if we have the entire system before us. This happens very infrequently—an example where it has happened is in the axiomatic of Euclidean Geometry (by Hilbert and others). For the most part we do not have all the possible propositions before us. Rather we are seeking the proper axioms from which we can derive the greatest quantity. A number of problems arise here. Given a set of axioms $A_1 \ldots A_n$, can any one, say A_p, be derived from the others? This is the question of *independence*.

The second question is, "Does there exist any general method by means of which we can determine the smallest set of axioms necessary and sufficient for the deduction of all our theorems?" If we have all our theorems before us (implying therefore that there is a finite number of them), then if we can derive all our theorems we answer the question affirmatively. If, however, we have an infinite number of possible theorems, or an unknown number of theorems, we face a very difficult problem. This has been called the problem of Categoricity or Completeness.[7] The problem is—how are we to determine the axioms necessary and sufficient for our axiom system? This must involve the possibility of deriving (or showing impossible to derive) any given meaningful theorem from our axioms.

Categoricity may be defined as follows: Given a postulate set $A_1 \ldots A_n$ and an arbitrary formula of the system. Then if the formula can be shown to be derivable from $A_1 \ldots A_n$ or inconsistent with $A_1 \ldots A_n$, $A_1 \ldots A_n$ is said to be categorical. If the formula and its contradictory can both be derived from $A_1 \ldots A_n$, it is said to be *inconsistent*—otherwise *consistent*.

[7] Cf. Fraenkel, p. 347. Also W. Dubislav, *Die Philosophie der Mathematik in der Gegenwart*, Berlin, 1932.

An attempted solution of the question of completeness for certain axiom sets (in particular the sets of Huntington and Kline on "betweenness") was given by C. H. Langford in an article entitled "Analytic Completeness of Sets of Postulates" (*Proc. London Math. Soc.*, Vol. 25 (1926), p. 115). The method he used is exhaustion of all element combinations and a proof that they are all either derivable from or incompatible with the axioms. Likewise, E. L. Post has offered a method for the solution of this problem in the propositional calculus by means of truth value matrices.[8]

Further light is thrown upon these criteria by consideration of the purposes of axioms. What are the axioms supposed to do? Returning to a consideration of the postulate set given previously, we notice that it begins with the assumption that there exists a class K, etc. This is evidently one of the purposes of our axioms—to assume the existence of certain entities.

It is to be noticed that some of these entities are defined systemically, while others are defined in terms of these. We might call these initial elements the *atoms* of the system. The first purpose of our axioms then must be to establish the existence of the atomic elements of our system. These axioms delimit the properties which the assumed entities are to have, and therefore delimit also the various fields of application or possible representations of the axioms. In other words, the axioms delimit the sphere of investigation, and indicate the "essence" of the objects being studied.

This is an important purpose of the axiom set as a whole—to define the area of applicability. Very frequently an axiom is inserted or removed with the express purpose of either increasing or delimiting the sphere of applicability. For example, Fraenkel[9] wishes to exclude those aggregates of the Mengenlehre that lead to paradoxes. In order to achieve this goal he includes another axiom in his set which limits the generality of the variables involved. It goes without saying that such a method must be used with great care.

The generality of the variables must be as complete as possible to insure that we are dealing with the essence of the object, and not a special case of that object. Therefore if an axiom is introduced which limits such generality, the system will lack generality, and hence may have to be rejected.

---

[8] E. L. Post, "Introduction to a General Theory of Elementary Propositions," *Amer. Jour. Math.*, Vol. 43 (1921), p. 163. Also H. Behmann, "Beiträge zur Algebra der Logik, etc.," *Math. Ann.*, Vol. 86 (1922), p. 163; L. Löwenheim, "Über Moglichkeiten im Relativ kalkül," *Math. Ann.*, Vol. 76 (1915), p. 447.

[9] *Einleitung in die Mengenlehre*, p. 247.

To illustrate the delimitation of spheres of applicability, suppose we have the following axioms:

1. There exist at least two elements a and b, $a \neq b$.

2. $a+b=b+a$

3. $a+0=a$

4. $(a<b)(b<c)$ implies $(a<c)$.

Without any further axioms there is a wide range of objects for which these could serve as a foundation for the development of formal properties. To mention only a few: the set of numbers $\{2, 3\}$; the set of classes, etc.

If now I add the further axiom

5. $a.a=a$

certain possible applications drop out. For example, I cannot use the set for numbers.

This is not to imply that there is one and only one application for every set of axioms. Nor does it imply that there must be an interpretation, although the existence of an interpretation shows the set to be consistent. In fact, the existence of an interpretation really means that there is an isomorphism between the set under consideration and some other set acknowledged (or assumed) to be consistent. There is also accepted the assumption that if a set is isomorphic with another which is consistent, then the given set itself is also consistent.

A postulate set (if consistent) describes the structure of a *possible* system of entities. And as Eaton says, "A postulate set is not dependent on any single interpretation, it deals with possibilities—and if it had no interpretation it would still describe the structure of a possible system though it would be of little interest to mathematics."[10]

The postulate set prescribes the universe of objects (or types of universes) to which it may be applicable. Axioms define the type of objects satisfying them, and cannot be merely rules of a game played with meaningless symbols. Dubislav[11] does admit that these "games enable men to use the objects of their environment."

Our axioms must, if they are to be of use, assume the existence of certain operations, as, e.g., addition, negation, etc. This is necessary in order to proceed. With entities alone nothing further can be done. But with entities and operations, we can derive new entities and then proceed to our consequences. Further, a set of relations among the elements must also be introduced in our axioms. The purposes of our

[10] Ralph M. Eaton, *General Logic*, 1931, p. 473.
[11] *Erkenntnis* 1.

axioms can be reduced to two general heads: (1) to postulate the existence of certain entities; (2) to define entities and relations.

Axioms are chosen, as we have seen, by a combination of the four methods noted above. They give us the "essential" properties from which the entire system can be derived. In so far as they do express the essence of the system they are recognized as indubitable and "self-evident" for that system. Only in so far as we are able to recognize the essence of the system can we retain the old notion of self-evidence of axioms. Until we have done so, no proposition—whether it be selected as axiom or not—is more self-evident than any other. The colorless term "primitive proposition" which is in use today in place of "axiom" was introduced because the choice of axioms was taken to be arbitrary. That this arbitrariness is illusory when we regard the essential nature of the system we are constructing is evident in the light of the criteria of acceptability of a set of axioms—especially the property of categoricity.

Furthermore, axioms introduce certain elements, relations and operations. They would not be accepted if they did not introduce elements appropriate to the system, related in ways proper to them. The recognition of the fact that these are the appropriate entities and they are related properly can only be the result of immediate recognition based on an understanding of the essence of the system. The fact that various sets of axioms can give rise to the same deductive system merely means that the essential properties of a system can be expressed in different ways. Only the confusion between the *ways of expressing* the essential properties and the properties expressed gives rise to a refusal to accept the self-evidence of axioms for a given system. But it must always be remembered that axioms are self-evident only relative to a given system.

At the bottom of every postulate set, except the logical set, is a rule or method of drawing inferences. This is taken from logic, hence it is felt to be unnecessary to write these down. The mathematician who is endeavoring to build up his system justly assumes that the logician has done his work properly, and he can therefore build on the results of the logical system. In other words, we have in every axiom set an implicit method of proof which Hilbert indicated by the schema

$$\begin{array}{c} p \\ p \rightarrow q \\ \hline q \end{array}$$

The simplicity of this schema is due to the fact that Hilbert, as we have seen earlier, assumes that Russellian logic is completely general and accept Russell's definition of the $\rightarrow$. p may denote the axioms, and q the conclusion or theorem.

NATURE AND PURPOSE OF THEOREMS

The axioms as we have discussed them are only a subset of all those propositions constituting the axiom system. The remaining set of propositions which can be derived or proved on the basis of the axioms are called the *theorems* of the system. The methods of proving theorems have been discussed earlier and we will not repeat what was said there.

A theorem is an explication of what is already involved in the axioms. It both specializes and particularizes the statements made by the axioms. A theorem draws in the detail of the postulational system; making explicit the structure of the system. In so far as it does that, it brings to the mind of the mathematician material that is psychologically new. The significance of the axioms and their relation to each other is thereby made more evident. So far as the system itself is concerned, if a proper set of axioms is obtained no theorems need be drawn. However, the individual is aided in detecting properties of the set of postulates by deriving theorems. Thus consistency is defined in terms of the derivation of both p and $\sim p$, categoricity in terms of the derivation of either p or $\sim p$ (but not both). Where methods of demonstrating these properties can be discovered which do not necessitate the actual proving of these theorems, the theorems become unnecessary.

A theorem is not involved in the axioms in the same way that the folds of a telescope are. They are not gotten by "pulling out the telescope." A proof makes evident the relation between the various *facts* of the system. In so far as propositions lead to all the facts, i.e., are related either directly or through other facts, they are selected as axioms. A theorem is therefore not merely the axioms in another form but also adds a new element of knowledge.

PROPERTIES OF POSTULATE SETS[12]

We wish now to turn our attention again to the properties of independence, consistency, and categoricity.

Consistency is defined as follows: A set of postulates is said to be consistent if it is impossible to derive two propositional symbols (or propositional function symbols) which are contradictory to each other.

Probably the only generally available method and the one used in the illustration at the beginning of this chapter is the method of *evaluation* or *interpretation*. Since the very form of two contradictories makes it impossible for them to be true together, the demonstration of the existence of values satisfying any two or more given propositions simultaneously would prove that they are not contradictory. If now we can

[12] Hilbert u. Ackermann, *Grundzüge d. theor. Logik*, 2nd ed. Cf. also my more detailed discussion in Kattsoff, "Postulational Methods III," *Philos. of Science*, Vol. 3, No. 3 (July, 1936), pp. 375–415; also for a more elementary introduction, Young, *Fund. Concepts of Algebra and Geometry*, New York, 1917, Chap. 5.

derive two contradictory propositions from a given set of axioms, this would mean that the contradiction is somehow involved in the axioms and they could not all be satisfied simultaneously by a given interpretation. Hence, if it is possible to find an interpretation, the axiom set will be known to be consistent, i.e., no contradiction is involved in them.

In most cases, arithmetic is used for the purposes of interpretation. This, of course, presupposes the consistency of arithmetic and hence makes the problem of the consistency of arithmetic an extremely important one.[13] For some systems a more direct proof of consistency is necessary.

However, where one fails to find an evaluation and cannot demonstrate an inconsistency, nothing is known.

Heyting defines the method of evaluation as follows: "By an evaluation of the formulae of a given calculus we mean a rule which assigns to each formula either the value t (for "true") or the value f (for "false") in such fashion that:

1. Every axiom has the value t.

2. From formulae having the value t we obtain by application of the rules of the calculus only t-formulae.

3. If the formula A has the value t, then $\sim A$ has the value f and conversely."[14]

Hilbert uses the values 0 and 1 instead of t and f.

Independence is defined as follows:

An axiom set is said to be *independent* if no axiom can be derived from the remaining axioms. Or more formally: An axiom set $A_n (n=1, 2, \ldots s)$ is said to be independent if it is false that $(A_n - A_i) \rightarrow A_i$, where $A_n - A_i$ denotes the postulate set which remains when A_i is withdrawn.

By logical considerations, this problem can be reduced to that of consistency and its solution referred to the consistency of another set of postulates.

A theorem in logic tells us that if $p \rightarrow q$, then $p . \sim q$ is impossible; and conversely if $p . \sim q$ is impossible, then $p \rightarrow q$. Furthermore, if $p . \sim q$ is impossible, then $p . \sim q . r . s$, etc., is impossible.

Now, let us suppose we have a postulate set p, q, r, s, t, u. The problem is (say) "Does $q . r . s . t . u \rightarrow p$.?" Using the logical principles if $q . r . s . t . u \rightarrow p$, then $q . r . s . t . u \sim p$ would be impossible, i.e., inconsistent. Therefore, if $q . r . s . t . u . \sim p$ can be shown consistent by evaluation or

[13] Cf. Gentzen, "Neue fassung des Widerspruchsfreiheitsbeweises für die reine Zahlentheorie," in *Forschungen zur Logik*, etc. Heft 4, Leipzig, 1928.

[14] Heyting, *Math. Grund.*, p. 44.

other methods, p would be independent of the other axioms. In this way if we can show the following sets consistent, the original set would be an independent set:

$\{p.q.r.s.t.\sim u\}$, $\{p.q.r.s.\sim t.u,\}$
$\{p.q.r.\sim s.t.u\}$, $\{p.q.\sim r.s.t.u,\}$
$\{p.\sim q.r.s.t.u\}$, $\{\sim p.q.r.s.t.u.\}$

In the preceding discussion no attempt was made to analyze more closely the definition of independence. In any set of propositions there are involved a set of concepts, a set of relations affirmed to exist between these concepts, and a set of properties. As a result, we must consider briefly what may be called "intensional independence." If two concepts are entirely unrelated, of course, the two propositions which involve them must be independent. On the other hand, if it can be shown that two concepts are derivative, one from the other, then the propositions which enunciate their properties must be derivative, and hence nonindependent.

For example, if we accept Whitehead and Russell's doctrine that the concept of implication is definable in terms of disjunction, i.e., implication can be derived from disjunction, then any theorem relating to implication must be derived or derivable from those of disjunction. In fact many of the theorems of *Principia* about implication are proved by reducing the implicative relations to disjunction. Only one of these two can be taken as a primitive idea. Two axioms are independent if a new concept occurs in one which is not in the other,—independent in the sense that one cannot be derived from the other.[15] On the basis of this it follows that two propositions, one of which contains a concept contradictory to a concept in the other, are independent, but *not*, of course, consistent.

Taking the set of elements in any set of propositions, we can have the following cases: (1) the set of propositions affirm of the same concepts,[16] the same relations; (2) they affirm of the same concepts different relations; (3) they affirm of the same concepts the same properties; (4) they affirm of the same concepts different properties; (5) they affirm of different concepts the same relations; (6) they affirm of different concepts different relations; (7) they affirm of different concepts the same properties; (8) they affirm of different concepts different properties.

Clearly in the cases (1) and (3) we do not have independent propositions. In so far as the same concepts must have the same properties,

[15] Cf. Weyl, *Handbuch der Phil. der Math.*, p. 18.
[16] We mean by concept here the subjectival concept as opposed to the adjectival and relational concept.

case (4) is impossible unless the propositions affirm different aspects—in which case they are equivalent to one proposition. For example, "the red ball is red" and "the red ball is a ball" are together equivalent to "the red ball is both red and a ball." The remaining cases must be separately considered.

To affirm of the same concepts different relations gives two cases: (a) the relations are independent; (b) the relations are derivative. In (b) we have the case discussed above for implications, and the propositions are therefore non-independent. In (a) we have what are called "external" relations and the propositions are independent. As an illustration: given the concepts a, b, and the relations $\sim$ and $+$, then the propositions

$$\sim\sim a = a$$

$$a + b = b + a$$

are independent. This takes care of (2).

Case (5) is seen to involve dependence immediately, since by generalizing the propositions we can derive all the given propositions. As an example: given

1. A set of concepts a, b, c, d
2. A set of relations R, R'
3. A set of statements

   aRb cRd

   $aR'b$ $cR'd$

By generalization we can write instead: given

1′. A set of variables x, y representing concepts

2′. A set of relations R, R'

3′. xRy

4′. $xR'y$

Then the statements in 3 become theorems by the specialization of 3′ and 4′.

Case (6) involves independence immediately.

Case (7) means either the properties enunciated are not all the properties, or else the concepts are not different. If the properties are not completely stated, it follows that the concepts are derivative under limitations.

Case (8) involves independence immediately.

To set up a logic of independence based upon intensional factors would, then, involve a consideration of the eight cases mentioned above. We do not intend, however, to undertake that task here. We are concerned with purely formal independence, which is a prior question.

Although we cannot say that if two concepts can be attributed to the same object they are derivative, we can say that if two concepts cannot be attributed to the same object they must be incompatible. Hence it follows that given two concepts, if we can attribute them to the same object, they are compatible. It follows, further, that in the ordinary proof of independence by interpretation we demonstrate at the same time the compatibility of the concepts involved.

The underlying assumption in interpretations which make an appeal to the experiential world is that contradictory attributes cannot exist in the same objects. It follows from this assumption that if we can give such an interpretation of our propositions we have been successful, because the propositions involved are compatible or consistent. This assumes the validity of the law of contradiction, and of course involves the metaphysical doctrine of Hegel, "The real is the rational."

Completeness or categoricity (Vollständigkeit) may be defined in a number of ways.[17]

A system is said to be complete

1. If when A is a formula, A is either demonstrable or inconsistent with the axioms.

2. A weaker form is given by Hilbert; if whenever A is consistent with the axioms, it is demonstrable.

3. If when a new formula, not derivable from the axioms, is added to the axioms, the resultant system is inconsistent.

The difficulties involved in the discussion of this property carry us beyond the scope of our interests. We shall conclude this chapter with an illustration of the demonstration of the three properties mentioned. We have briefly indicated demonstrations of independence and consistency in the illustration at the beginning of this chapter, but repetition may make the matter clearer.

We select the axioms used by Hilbert (p. 23 of 2nd ed. of Hilbert-Ackermann, *Grundzüge*, etc.)

a) $x \vee x \rightarrow x$

b) $x \rightarrow x \vee y$

---

[17] Heyting, p. 48.

c) $x \vee y \rightarrow y \vee x$

d) $(x \rightarrow y) \rightarrow [z \vee x \rightarrow z \vee y]$

(where $\vee$ indicates disjunction, usually denoted by $+$, and $\rightarrow$ indicates implication, which is defined as $x \rightarrow y = \text{df. } \bar{x} \vee y$).

CONSISTENCY (Hilbert, pp. 32–33)

The symbols x, y, and z are to be considered arithmetic variables capable of assuming only the values 0 and 1. $x \vee y$ shall denote the arithmetic product. $\bar{x}$ shall be 0 when x is 1 and 1 when x is 0.

Now we can see that $\bar{x} \vee x$ is always 0 since $\bar{x} \vee x$ denotes the arithmetic product:

if $x=1$, then $\bar{x}=0$ and $\bar{x} \vee x = 0 \times 1 = 0$;
if $x=0$, then $\bar{x}=1$ and $\bar{x} \vee x = 1 \times 0 = 0$;

Axiom (a)

$x \vee x \rightarrow x$ is by definition $\overline{(x \vee x)} \vee x$;
if $x=1$, then $\overline{(1 \times 1)} \times 1 = 0 \times 1 = 0$;
if $x=0$, then $\overline{(0 \times 0)} \times 0 = 1 \times 0 = 0$.

Axiom (a) *therefore is always 0.*

Axiom (b)

$x \rightarrow x \vee y$ is by def. $\bar{x} \vee (x \vee y)$;
if $x=1$, then $\bar{x}=0$ and no matter what y is, the product is 0;
if $x=0$ then in $(x \vee y)$ we always get 0 and the product is 0;

Axiom (b) *therefore always has a value* 0.

Axiom (c)

$x \vee y \rightarrow y \vee x$ is $\overline{(x \vee y)} \vee (y \vee x)$;
if x is 0 the $(y \vee x)$ will always result in 0 and cause the product to be 0;
if y is 0 the same thing happens;
if x is 1 and y is 1, then $\overline{(x \vee y)} = \overline{1 \times 1} = \bar{1} = 0$ and the entire product is 0;

Axiom (c) *is therefore always 0.*

Axiom (d)

$(x\to y)\to[z\vee x\to z\vee y]$ is $\overline{x\to y}\vee[z\vee x\to z\vee y]$ which is $(\overline{\bar{x}\vee y})\vee[\overline{z\vee x}\vee z\vee y]$;

if $z=0$, then $z\vee y$ is 0, making the entire product 0;

if $y=0$, the same thing happens.

This leaves only the cases (1) $x=0, y=1, z=1$, and (2) $x=1, y=1, z=1$.

For case (1)

$\overline{\bar{x}\vee y}=\overline{\bar{0}\times 1}=\overline{1\times 1}=\bar{1}=0$ and the product is 0.

For case (2) the second factor gives

$\overline{z\vee x}=\overline{1\times 1}=\bar{1}=0.$

Axiom (d) *is therefore always 0.*

On our interpretation all axioms have the same value 0 and by definition are consistent.

INDEPENDENCE (Hilbert, pp. 33–34)

We shall prove only the independence of (b).

x, y, z are variables capable of taking on values 0, 1, 2.

$\vee$ is defined as follows: ($x\vee y$ is same as $y\vee x$)

x	0	1	2
0	0	0	0
1	0	1	1
2	0	1	1

and x is defined as $\bar{0}=1$, $\bar{1}=0$, $\bar{2}=2$.

Now evaluating the axioms, when $x=2$ and $y=1$ we get

Axiom (a) $\overline{x\vee x}\vee x$

for $x=2$, we get

$\overline{2\vee 2}\vee 2=\bar{1}\vee 2=0\vee 2=0$

Axiom (b) $\bar{x}\vee(x\vee y)$

for $x=2, y=1$, we get

$\bar{2}\vee(2\vee 1)=\bar{2}\vee 1=2\vee 1=1$

(this is the evaluation which shows (b) independent).

Axiom (c) $(\overline{x \vee y}) \vee (y \vee x)$
for $x=2, y=1$, we get
$(\overline{2 \vee 1}) \vee (1 \vee 2) = \bar{1} \vee 1 = 0$

Axiom (d) $\overline{\bar{x} \vee y} \vee (\overline{z \vee x} \vee z \vee y)$
for $x=2, y=1$, we get
$\overline{\bar{2} \vee 1} \vee (\overline{z \vee 2} \vee z \vee 1)$
since $\overline{2 \vee 1} = \bar{1} = 0$, the product is always 0.

Hence, of the axioms for $x=2, y=1$,
a) evaluates to 0
b) evaluates to 1
c) evaluates to 0
d) evaluates to 0

Axiom (b) is therefore independent of the other axioms.

COMPLETENESS (Hilbert, p. 35)

Hilbert proves completeness in accordance with the third definition given above—the addition of the nondemonstrated formula leads to contradictions.

Let *A* be any formula nonderivable from the axioms. We transform it into the form of a conjunction of disjunctions, using the identity $\overline{\bar{x}.y} = x \vee y$, as was done when we illustrated consistency and independence proofs in the immediately preceding pages. Let *B* be this form. "Since *B* cannot be derived either, there must occur among the disjunct-factors (Summanden) of *B* a simple product *C* in which no two parts are opposites. If in *C* we replace each negated symbol $\bar{x}$ by x we get a disjunction of the form $x \vee x \vee x \vee \ldots \vee x$, which by the rules of the propositional calculus is equivalent to x. If we assume *A* to be valid, then *B* and *C* and hence x are also. Now then we need but replace x by $\bar{x}$ also and we get a contradiction."

CHAPTER SIXTEEN

MATHEMATICS AND REALITY[1]

Somewhere Kelvin has said that "Mathematics is the only true metaphysics." Certainly for many years we have heard it repeated that true knowledge of nature is possible only when we have reduced nature to mathematical form, and the progress of mathematical physics has aided the spread of this idea. From the Pythagoreans through Plato, Descartes, and Eddington, mathematics has been conceived as in some way essentially related to reality. Even those who, like Carnap, insist that questions about reality are meaningless still tell us that is is possible to set up a correspondence between symbol and experience in such a way that the symbol may be used in place of the experience. Yet so much of our analysis of mathematics seems to be as far from experience as one can imagine!

The development of the number system seemed to be a purely rational one necessitated not by experience of reality, but by the logical need of the principle of no exception. Even the definition of number was, with some exceptions, either taken as given or defined in the terms of logic. The discussion of the foundations of mathematics, with the exception of Pasch's ideas, concerns to a great degree logical problems. And logic would, of all sciences, certainly not be derived from reality by empirical methods! Or so it would seem.

Yet, even the problem of the relation between logic and reality has not been completely answered. If mathematics were reducible to logic completely, then our problem would become the problem of the relation between logic and reality. But it is possible to analyze our problem without discussion of this more general one.

The two extreme positions that can be taken concerning the relation between mathematics and reality are, (1) there is no relation—i.e., a complete dualism, and (2) mathematics is the basis of reality. The first is the position taken by the logical positivists, and in a modified

[1] The literature on this is extensive. Only a very few sources are given: Ernst Cassirer, *Substance and Function;* J. B. Shaw, *Lectures on Phil. of Mathematics;* L. Brunschwig, *Les Étapes de la Phil. Math.*, Paris, 1929; F. Gonseth, *Les Math. et la Réalité*, Paris, 1936; F. Maugé, *L'Esprit et le Réel*, Paris, 1936; also the sections in W. Dubislav, *Die Phil. d. Math.*, and Heyting, *Math. Grund.;* J. Baumann, *Die Lehren von Raum, Zeit, u. Mathematik*, Berlin, 1869; Hans Hahn, *Logique, Math. et Conn. de la Réalité*, Paris, 1935; *Logik, Mathematik und Naturerkennen*, Vienna, 1933.

form by Plato and Kant. The second position was taken by the Pythagoreans, Galileo, Kepler, Descartes, Eddington, and Jeans. In a modified form it is also the position of most mathematical physicists, the phenomenological school of Husserl, Hegel, and others.

The positivists base their claim on their interpretation of the nature of mathematics. For the positivists, mathematics is a branch of logic, and logic is merely a language, all of whose statements are tautologies. In so far as mathematics is a language, it is a construct of symbols formed by the investigator for the purpose of preserving his empirical data. The selection of symbols used is made purely arbitrarily. The fundamental statements of an *applied* language express facts which can be empirically verified. So far as a pure language is concerned, the fundamental statements merely express the rules of combination of the symbols. All the statements of mathematics are true, then, because they are derived in accordance with the rules of the language. They are true because of their form and are called tautologies.

In so far as this is so, argue the positivists, the laws of mathematics tell us nothing about objects as such. For example, $2+2=4$ is absolutely true because given the meaning of 2, 4, +, =, it will be seen that $2+2$ expresses the same thing as 4. But this does not tell us anything about objects. Or, to take another illustration (but from logic, the law of excluded middle applied to a specific entity) say, this paper is either white or not-white. This does not tell us any more about this paper than we knew before. It is a true statement because of its form alone.

In brief, for the positivists, mathematics is the science of the forms of our *expression*, as for Kant it was the science of the forms of our *intuition*. But the positivists, as Kant, face an unanswerable problem on their own bases. How, then, does mathematics enable us to control events—to predict events—and to get knowledge of what is the case? How can an expression which is arbitrary be applied? In brief, the question reduces itself to the problem: How is applied mathematics at all possible? Kant had made experience possible by the recognition of the forms of experiencing, but he never could make it actual nor explain why it came about. The positivists do not even make mathematics possible. A positivistic, i.e., syntactical, analysis of mathematics is important but presupposes the structure of mathematics to be established and given. In the natural sciences it is possible to leave the construction of the science to the scientists and no one would confuse the syntax of physics with physics—so in mathematics the syntax of mathematics is not to be confused with mathematics.

The possibility of setting up an isomorphism between a mathematical structure and a sphere of reality is explicable only on the grounds of common structural properties. If we recall the discussion of the selection of axioms in the preceding chapter, it will be remem-

bered that it was said there that experience suggests our axioms. It is only a suggestion because a process, not of mere abstraction but of the recognition of ontological properties, is necessary. It is the "essence" of the "pure object" that must be seen before mathematics is possible. Mathematics then would express properties of objects which are universal and necessary in the sense that they belong to the pure object.

Of course, having established, or recognized, the bases of a system, we can develop the properties involved in the system itself. Thus the various foundation theories are developed apparently independent of reality. But that mathematics has some structural properties in common with reality is evidenced by every discovery predicted by mathematical physics. This is not refuted by arguing that experience *might* not have verified the prediction. The point is that it did! But it is not strange that those concerned with the problems of mathematics as such, would be agreed that mathematics had nothing to say about objects.

For Kant, too, mathematics could tell us nothing about the object but only about the forms of perception. The thing-in-itself remains unknowable. But so far as the object-as-perceived is concerned, it is given to us in these forms which mathematics studies. Hence, mathematics for Kant does have a definite relation to the experiential realm. It gives us the properties of objects-in-perception. In so far as it does give us these properties, mathematics is a science. Hence, for Kant a body of knowledge becomes a science only to the extent it is mathematical.

The formalists insisted that every branch of learning be in the form of an axiomatic system. They would agree with the positivists and with Kant that mathematics tells us nothing about objects. But their reason for that fact is more akin to that of the positivists. For formalism, mathematics is a pure calculus set up and played according to pre-established rules. As a calculus it can be applied to objects, but is not indicative of any properties of the object.

Historically speaking, of course, the formalists are wrong. Man did not invent a language, or a calculus, and then apply it to objects. We do not find pure mathematics among primitives. Nor did algebra, geometry, arithmetic, calculus, etc., arise by someone's constructing these theories as games with rules. Quite the contrary—they arose as abstractions from concrete applications. For primitive people, number is a physical quality of groups of objects.[2] As a matter of fact, the remnants of this idea are evident in Frege and in Russell's definition of number as a property of classes. This intimate connection between

[2] Gonseth calls logic "La physique de l'objet quelconque." For discussion, *see* F. Gonseth, *Les Mathématiques et la Réalité*, Chap. 8.

mathematics and reality is lost sight of by studying the *logical* development of the number system as we did. What has happened has been pointed out by Brouwer and Heyting—a confusion between the means of expressing mathematical truths and the mathematics itself.

This does not mean that mathematics is an empirical science obtained by the inductive procedures of natural science based on observation by means of our senses. But it does mean that mathematics is empirical in the sense that it arises through the observation of "essences" (in Husserl's sense), by means of what Husserl calls *essential intuition.* The probability character of the natural sciences is due to the use of sensory experience. The categorical character of mathematics is due as much to the use of essential experience as to the tautological character of the system.

If the system of mathematics were based on arbitrary assumptions, the entire structure would become arbitrary. The positivists and formalists would then be right. But it is extremely difficult to see how an arbitrary system could have any applications whatsoever. If there is any arbitrariness it is not in the axioms, but rather in the particular type of symbol selected to be used in writing. Whether we use V or 5 may be arbitrary, but what the fundamental signs *mean* cannot be. But the change from a notation I, II, III, IV, V, VI, etc., to 1, 2, 3, 4, 5, 6, etc., itself shows that even here arbitrariness has its limits. A great deal of the argument for the arbitrary character of mathematics is due to the conventional aspects of all symbols. This is especially true in Poincaré's work.

Pure mathematics, such as has been discussed, really is the study of mathematics with some of its aspects suspended. Applied mathematics is the consideration of those suspended aspects. The true nature of mathematics can be found in a consideration of both aspects. To make of the two aspects two different, separate fields is to introduce a sort of schizophrenia into the scientist. Such a schizoid point of view is as common to many scientists and scholars today with respect to their field of work and its application as it was to the scholar of yesterday with respect to reason and religion.

This point of view is frequently expressed as a desire on the part of the scientist to seek truth for truth's sake. But it must be remembered that although we speak of "Truth for Truth's Sake," actually, from the nature of truth itself, the implications of this statement are impossible. To attempt to develop the theory of mathematics in itself is a valid enterprise, but the justification of mathematical study lies in the applicability to the whole realm of knowledge. The mathematician who said, "Thank God no one has yet found an application of my work," was really thanking God that it was still possible that his work was false. Truth is a property of systematic wholes.

In the last analysis, then, a theory is true if its relations with the rest of reality are definitely established. This is best done by showing its applicability to at least one other realm. It is a well known fact that a system is most easily shown to be consistent if its terms can be interpreted. Ultimately, the consistency of sets of postulates (i.e., systems) is referred to the rationality (consistency) of reality.[3] The true significance of mathematics is grasped only in its relations to the rest of nature.

There is, however, another meaning we can give to the statement "Truth for Truth's Sake." Recognizing that truth is a function of systems, i.e., that anything is said to be true if it fits into its proper system, it follows that if we have truth, we have system. And if we have the system, we can use it without destroying it. It follows, then, that where we have truth, we can apply it, i.e., truth implies workability or control. This is the kernel of truth in pragmatism and positivism.[4] Whenever a proposition is true, it can be verified in some way. The nature of this "verification" becomes an important problem. This means that the pure mathematician can acquire pure theorems even if he does not actually apply these propositions. The *possibility* of application is there. It is in this fact that we find the explanation of the applicability of mathematics to the external world. The problem is really one of our own creation. It arises only if we attempt to set up a dualism between mathematics and reality. This dualism once set up is unbridgeable. It is only when we recognize that each is essentially an aspect of the other that we transcend the dualism.

"Truth for Truth's Sake" may then mean the discovery of concrete truth—the search for truth which leads us to the power to control our world and advance civilization. As a consequence of this point of view, it follows that any new development in mathematics needs but the passage of time before it finds its application. Not only that, but we are certain that if our work is logical, there *must* be a field of application. This is as true of Mengenlehre and the theory of Boolean Algebras as it is of Geometry and the Calculus.[5] (Reality as used here may be taken, in a limited sense, to denote the class of all entities which are the subject of investigation by the natural scientist.)

If this position is accepted, then a number of important consequences follow: (1) Since reality is a related, ordered system, all

[3] Cf. Kattsoff, "Postulational Methods III," *Phil. of Science*, Vol. 3, pp. 375 ff.

[4] Cf. Hahn, *Logique Mathématiques* (trad. Vouillemin), Paris, 1935, p. 47. Also Edgar Zilsel, "Moritz Schlick," *Naturwissenschaft*, Vol. 25 (March 12, 1937), p. 163.

[5] Cf. B. Croce, *Logic*, trans., Ainslie, New York, 1917, p. 362 ff. for the opposite view. Readers interested in the author's views concerning the nature of reaists are referred to L. O. Kattsoff, "Physics and Reality," *Philosophy and Phenomenological Resarch*, Vol. 5, No. 1 (1944), pp. 108 ff., and L. O. Kattsoff, "The Event in Rely a Ontological Unit," *Philosophical Review*, Vol. 55, No. 2 (1946), pp. 174 ff.

branches of mathematics are related and ordered. (2) Mathematics and logic are related in some way. (3) Mathematics is not a *creation* of the mind but a discovery by the mind of the relation between things.

Since this is not a treatise on metaphysics we shall not discuss in detail the nature of the relatedness of reality. Whether it is so related and ordered intrinsically or extrinsically, for mathematics to be applicable—for mathematics to be possible, an order must exist. As a matter of fact, as we have seen, mathematics has been defined as the science of order. But it is important to note that the existence of a relation between mathematics and logic does not of itself restrict the relation to one of identity or one of part-whole.

There are then two aspects in the verification of a mathematical theorem: (1) its derivability, (2) its applicability. Its derivability is not separate and distinct from its applicability. That is to say, the analytic character (i.e., tautologous character) of mathematics indicates a corresponding analytic character in reality.

Wittgenstein has pointed out (and positivists today accept) the doctrine that there are two types of propositions: (1) tautological, (2) empirical. Tautological propositions are of the general form "*p* is true or false." Such propositions occur in mathematics and logic. They are merely formal alterations in the fundamental assemblage of symbols. Their truth or falsity is determined analytically. These propositions say nothing about matters-of-fact. Nontautological (empirical) propositions make statements about actual matters-of-fact. They indicate something beyond themselves.[6]

From our point of view every *proposition* has both these aspects. Only when a proposition can be shown to be tautological and verified nontautologically (e.g., by describing a possible physical process of observation) can that proposition be permitted to become an integral part of our body of *knowledge*. This means that not only must we have a theoretical structure of mathematics, but also some field of application of that structure.[7] Although it is true that every proposition tautological with respect to a set of given data (i.e., every proposition which can be deduced from the set of *given* data) can be interpreted and applied, it does not follow that every nontautological proposition can be considered as tautological (i.e., can be deduced from the given data).

A tautological proposition can be applied by establishing a (1,1) correspondence between the symbols or groups of symbols in the proposition (if the proposition is an axiom) or the symbols in the set of axioms from which the proposition is derived and actual entities. That is, if

[6] Cf. J. R. Weinberg, *An Examination of Logical Positivism*, p. 58.
[7] Cf. Kaufmann, *Das Unendliche in der Mathematik*, Leipzig und Wien, 1930, p. 7.

the proposition given is $p=f(x,y,z,\ldots)$ where $x,y,z,\ldots$ are symbols which occur in p, then there are two cases:

1. *p is an axiom*—in which case we set up a (1,1) correspondence of $\begin{Bmatrix} x,y,z,\ldots \\ E_1E_2E_3\ldots \end{Bmatrix}$ where E_i $(i=1\ldots n)$ are given events or entities.

2. *p is a theorem*—then $x,y,z,\ldots$ are the symbols which occur in the set A_i of axioms and we can set up the same set of correspondences as in the Case 1, or else some of set $x,y,z,\ldots$ are compounds of symbols in A_i. In this latter case, our correspondence is set up between the symbols in A_i and E_i. This means that tautological propositions can be transformed by a given set of transformations into empirical propositions.

The converse operation—to transform an empirical proposition into a tautological one—would entail the difficulties of determining whether the events are compound or not, and then to determine the original set of propositions involving the *undefined* concepts. That is the problem of induction.

From what has been said, it follows that there is no real distinction between *pure* and *applied* mathematics. Pure mathematics is pure only in so far as it can be applied, and applied mathematics can be applied only as it is pure. If we wish to keep such a distinction we might say that the pure mathematician is interested in mathematical propositions as tautologies, the applied mathematician in the nontautological aspect of these same propositions. In terms of Russell's distinction between pure and applied mathematics, the variables (mathematical symbols) in the system must have some actual entities satisfying the system. (It appears to me that when Russell says, "In applied mathematics, results which have been shown by pure mathematics to follow from some hypothesis as to the variables are actually asserted, of some constant satisfying the hypothesis in question,"[8] he makes a tacit assumption concerning the relation between logic and reality which is equivalent to the point I am making. This assumption is: if a set of constants, i.e., symbols representing actual entities, satisfies the set of axioms of a mathematical, i.e., logical system, and thereby itself forms a *state of affairs*, then this *state of affairs* will give rise to, or cause, subsequent states which will verify the theorems derived from the axioms.) The pure mathematician seeks to set up the structure of mathematics as an ordered system. To accomplish this, he must travel in two directions. He must continue to discover new mathematical facts and he must continue to re-work the system to make it a coherent whole. "Science cannot proceed unless it ever glances back and reconsiders its structure

---

[8] B. Russell, *Principles of Mathematics*, Cambridge, 1903, p. 8 (reprinted in 1937 by W. W. Norton Co.)

in the light of new understanding and from a new point of view."[9]

One way of avoiding the difficulty in applying a mathematical system to reality is to consider the system as a kind of function—to use Keyser's term, a Doctrinal Function. Then, it is claimed, all one need do is to assign constants to the variables. This does not answer the problem but either avoids it or begs it. In ordinary mathematical analysis a function arises by generalization from concrete cases. If this is what is meant, then clearly a doctrinal function arises also in this way, and hence has a definite relation to reality. The process of specialization can bring us back to the cases from which it arose. At any rate, a generalization retains specific properties of the cases from which it is made.

If this is not meant, then the only alternative seems to be a type of Leibnizian pre-established harmony.[10] In technical terms, the isomorphism between the symbols of the system and reality would become a fortunate accident or divinely created state of affairs. It would therefore appear that any appeal to some sort of isomorphism between the symbols of the system and reality would become a fortunate accident or divinely created state of affairs. It would therefore appear that any appeal to some sort of isomorphism would raise problems greater than the one avoided. But even granting isomorphism as a solution, this would prove our point.

Two systems are said to be isomorphic if they can be put into (1,1) correspondence. This really means that the two systems have the same structure, since not only would the correspondence be between entities but also between the relations among the entities. Therefore, if reality and mathematics are isomorphic they have the same structure. Isomorphism is possible because there is identity of structure.

This point is not destroyed but rather substantiated by the example of Euclidean, Non-Euclidean geometry. The difficulty is cleared up by the very fact that *a* geometry is necessary for experience. We know that all attempts to determine whether actual (or, better, scientific) space is Euclidean or Non-Euclidean have been unsuccessful. We also know that Poincaré said the question was meaningless and which geometry we used was a matter of convention. Newton used Euclidean geometry and Einstein a Non-Euclidean geometry. But there exists an isomorphism which enables us to translate every Euclidean proposition into a Non-Euclidean one by a proper transformation. This is possible because the structure of both types of geometry is the same. Not merely that, but the ability to use geometry in physics which, as

---

[9] M. Pasch, *Grundlagen der Analysis*, Leipzig und Berlin, 1909, p. iii.

[10] Cf. Hugo Bergmann, "Zur Geschichte und Kritik der isomorphen Abbildung," *Actes du Congrès Int. de Philos. Scientifique*, Vol. 7, Paris, 1936, pp. 65–66.

Lenzen points out,[11] is really a reduction of geometry to physics indicates an essential isomorphism between geometry and nature. And this isomorphism denotes an identity of structure.

In so far, then, as there is an identity of structure, reality has the same structural properties as does mathematics. It is this fact which leads Jeans to consider God—the Creator—as a mathematician. There are, therefore, many implications concerning the nature of reality that can be drawn from the consideration of mathematics. For one thing, mathematics is systematic, hence reality is also. The system of mathematics takes the form of a postulate system. Hence, the ontological aspects of postulate systems will give us the form of reality as a system.[12] To the various aspects of a postulational system there correspond definite aspects in reality. A realization of this is evident in consistency demonstrations by exemplification. Consistency is often demonstrated by finding a set of objects satisfying the axioms in question. This, then, means that the axioms may co-exist in a real sphere. Inconsistency means, therefore, that the axioms cannot co-exist, while independence may denote non-necessary co-existence.

The various parts of this correspondence need not be developed here. We wish merely to indicate the possibility of a metaphysics based on the applicability of mathematics to reality. Such a metaphysics would clearly be a return to a rationalism of the Cartesian type. This tendency is indicated in Eddington and reflected by Korzybski when he says "The only usefulness of a map or a language depends upon the similarity of structure between the empirical world and the map languages."[13]

The idea of system is probably the most dominant one in contemporary thought. We find it not merely in postulate sets, but in the *field* of physics—as well as in statistical physics—and in the *organicism* of biology, in the *Gestalt* and *topological* field of psychology, in *national self-sufficiency* of fascism, in the *region* of sociology, etc.

Mathematics and reality are closely related, even though we have come a great distance from the initial steps of counting and measurement. That this has been realized more or less clearly is evidenced by the influence of mathematics on philosophic thought from the days of Plato to the present era. Speculatively, the philosophy of mathematics has served as the core of many a philosophic debate. Long the last stronghold of a type of rationalism against the increasing triumphs of empiricism, mathematics at last seemed to have been taken by the

---

[11] V. F. Lenzen, "Physical Geometry," *Amer. Math. Monthly*, July, 1939, pp. 324–34.

[12] L. O. Kattsoff, "The Relation of Science to Philosophy in the Light of Husserl's Thought," *Philosophical Essays in Memory of E. Husserl*, Harvard Press, 1941.

[13] A. Korzybski, *Science and Sanity*, 1933, p. 61.

empiricists—but not by demonstrating the empirical bases of mathematics so much as by trying to show mathematics to be merely a language constructed arbitrarily.

This reduction, it now becomes evident, does not prove the point—since any language has empirical reference. Mathematics and its application furnish an argument neither for a pure empiricism nor for a pure rationalism. It rather shows the close interrelation of the two in a unique fashion. The development of mathematical theory and the investigation of the roots of mathematics have shown the complementarity of both approaches to each other. Critical and speculative philosophy of mathematics and pure and applied mathematics are both aspects and not separated spheres.

The concept implied here is a dialectical interpretation of the relation of mathematics to reality. They are not to be considered to be separated realms but rather two aspects, to know either of which is to be led to the other. Mathematics may be abstracted and dealt with deductively as a static body of propositions, but to understand it as a phenomenon it is necessary to treat it dynamically—as leading to other fields. The same thing holds true of the study of reality. Mathematics is not merely a body of symbols, but is a construct of humans which is evolved in a society and which has relation to empirical events. It, therefore, has syntactical, sociological, psychological, and scientific fields of investigation. To confuse or to isolate syntactical, pragmatic, and semantic problems in mathematics always leads either to confusion or to abstraction.

BIBLIOGRAPHY

ACKERMAN, W. "Begründung des 'tertium non datur' mittels der Hilbertschen Theorie der Widerspruchsfreiheit," *Mathematische Annalen,* Vol. 93 (1925), pp. 1–36.

AMBROSE, A. "Finitism in Mathematics," *Mind,* Vol. 44 (1935), pp. 186–203; 317–340.

ANDRÉ, D. *Des Notations Mathematiques.* Paris, 1909.

AYER, R. *Language Truth and Logic.* New York, 1936.

BAUMANN, J. *Die Lehren von Raum, Zeit, u. Mathematik.* Berlin, 1869.

BECKER, O. "Mathematische Existenz," *Jahrbuch für Philosophie und Phänomenologische Forschung,* Vol. 8 (1927), pp. 439–809.

———. "Zur Logic der Modalitaten," *ibid.,* Vol. 11 (1930), pp. 496–548.

BEHMANN, H. "Beiträge zur Algebra der Logik," *Math. Annalen,* Vol. 86 (1922), pp. 163–229.

BELL, E. T. "On Proofs by Mathematical Induction," *American Mathematical Monthly,* Vol. 27 (1920), pp. 413–15.

———. "Place of Rigor in Mathematics," *ibid.,* Vol. 41 (1934), pp. 599–607.

BENNETT, A. A., and BAYLIS, C. A. *Formal Logic.* New York, 1939.

BENTLEY, A. F. *A Linguistic Analysis of Mathematics.* Indiana, 1932.

BERGMANN, H. *Bolzanos Beiträge zur Philosophischen Grundlegung der Mathematik.* Halle, 1909.

———. "Zur Geschichte und Kritik der isomorphen Abbildung," *Actes du Congrès Int. de Philos. Scientifique,* Vol. 7. Paris, 1936.

BERNAYS, P. "Axiomatische Untersuchungen des Aussagen-kalküls der Principia Mathematica," *Math. Zeitschrift,* Vol. 25 (1926), pp. 305–20.

BERNSTEIN, F. "Die Mengenlehre Georg Cantors und der Finitismus," *Jahresbericht der Deutschen Mathematiker-Vereinigung,* Vol. 28 (1919), pp. 63–78.

BETSCH, C. *Fiktionen in der Mathematik.* Stuttgart, 1926.

BIRKMEISTER. *Uber den Bildüngswert der Mathematik.* Leipzig, 1923.

BLACK, M. *Nature of Mathematics.* London, 1933.

BÔCHER, M. "Fundamental Conceptions and Methods of Mathematics," *Bulletin American Mathematical Society,* Vol. 2 (1904), pp. 115–35.

———. *Introduction to Higher Algebra.* New York, 1929.

BOLZANO, B. *Paradoxien des Unendlichen* (1851). Leipzig, 1920.

BROAD, C. D. "Critical and Speculative Philosophy," *Contemporary British Philosophy* (First Series). London, 1924.

BROUWER, L. E. J. "Intuitionism and Formalism," *Bulletin American Mathematical Society,* Vol. 20 (1913, pp. 81–96.

———. "Zur Begründung der intuitionistische Mathematik, I," *Math. Annalen,* Vol. 93 (1924), pp. 244–57; Vol. 95:II (1925), pp. 453–72.

———. "Uber die Bedeutung des Sätzes vom ausgeschlossenen Dritten in der Mathematik," *Journal f. d. reine und angewandte mathematik,* Vol. 154 (1925), pp. 1–7.

———. "Intuitionistische Betrachtungen über den Formalismus," *Sitzungsb. d. Preuss. Akad. d. Wiss. Phys.—Math. Klasse,* 1928.

BRUNSCHWIG, L. *Les Etapes de la Philosophie Mathématique.* Paris, 1912 (2nd edition 1922).

BURTT, E. A. *Metaphysical Foundations of Modern Physical Science.* New York, 1932.

CANTOR, G. *Transfinite Numbers.* Chicago and London, 1915.
CARNAP, RUDOLF. *Abriss der Logistik.* Vienna, 1929.
————. "Die Logizistische Grundlegung der Mathematik," *Erkenntnis,* Vol. 2 (1931), pp. 91–105.
————. "Die Antinomien und die Unvollstandigkeit der Mathematik," *Monatshefte für Mathematik u Physik,* Vol. 41 (1934), pp. 263–84.
————. *Philosophy and Logical Syntax.* Psychological Monographs. London, 1935.
————. *Logical Syntax of Language.* New York, 1937.
————. "Foundations of Logic and Mathematics," *International Encyclopedia of Unified Sciences,* Vol. 1, No. 2 and No. 3. Chicago, 1939.
————. *Introduction to Semantics.* Cambridge, Massachusetts, 1943.
————. *Meaning and Necessity.* Chicago, 1947.
CASSIRER, E. *Philosophie der Symbolischen Formen,* Vol. 3. Berlin, 1923–31.
————. *Substance and Function.* Chicago and London, 1923.
CHURCH, A. "Alternatives to Zermelo's Axiom," *Transactions of the American Mathematical Society,* Vol. 29 (1927), pp. 178–208.
————. "A Set of Postulates for the Foundation of Logic," *Annals of Mathematics,* Second Series, Vol. 33 (1932), pp. 346–66 (cited as *F. L.* 1). Second Paper—Vol. 34 (1933), pp. 839–64 (cited as *F. L.* 2).
————. "The Richard Paradox," *American Mathematical Monthly,* Vol. 41 (1934), pp. 356–61.
————. *Mathematical Logic* (Mimeographed Lecture Notes), Princeton, N. J. October 1935–January 1936 (cited as *M. L.*).
————. "An Unsolvable Problem of Elementary Number Theory," *American Journal of Mathematics,* Vol. 58 (1936), pp. 345–63.
———— (with J. B. ROSSER). "Some Properties of Conversion," *Transactions of the American Mathematical Society,* Vol. 39, No. 3 (1936), pp. 472–82.
————. "Introduction to Mathematical Logic," Part 1, *Annals of Mathematics Studies,* No. 13. Princeton, N. J., 1944.
————. Review of H. R. Smart's "Frege's Logic," *Journal of Symbolic Logic,* Vol. 10 (1945), pp. 101–2.
CHWISTEK, LEON. "Pluralité des Réalities," *Atti del V Congresso Internazionale di Filosofia.* Naples, 1924.
————. "Sur les fondementes de la logique moderne," *Atti del V Congresso Internazionale di Filosofia.* Naples, 1924, pp. 24–28.
————. "The Theory of Constructive Types," *Annales de la Société Polonaise de Mathématique,* Vol. 2 (1924), pp. 9–48; Vol. 3 (1925), pp. 92–141. (Reprinted by University Press at Cracow, 1925.)
————. "Neue Grundlagen der Logik und Mathematik," *Math. Zeitschrift,* Vol. 30 (1929), pp. 704–24.
————. "Une méthode métamathématique d'analyse," *Comptes-Rendus du Premier Congrès des Math. des Pays Slaves.* Warsaw, 1929, pp. 254–63.
————. "Die nominalistische grundlegung der Mathematik," *Erkenntnis,* Vol. 3 (1933), pp. 367–88.
————. "La Semantique rationelle et ses applications," *Travaux du IX Congres International de Phil. VI,* Logique at Mathématiques. Paris, 1937.
———— (with W. HETPER). "New Foundations of Formal Metamathematics," *Journal of Symbolic Logic,* Vol. 3 (1938), pp. 1 ff.
COHEN, M. R., and NAGLE, E. *Introduction to Logic and Scientific Method.* New York, 1934.
COMTE, A. *The Positive Philosophy* (trans. abridged by H. Martineau). New York, 1854.
COOLEY, J. C. *Primer of Formal Logic.* New York, 1942.
COURANT, R., and ROBBINS, H. *What Is Mathematics?* London and New York, 1941.
COUTURAT, L. *De l'infini mathématique.* Paris, 1896.
————. "The Principles of Logic," *Encyclopedia of the Philosophical Sciences,* Vol. 1, Logic. London, 1913, pp. 136–98.
CROCE, B. *Logic* (trans. Ainslie). New York, 1917.

CURRY, H. B. "Grundlagen der kombinatorischen Logik, Teil I," *American Journal of Mathematics,* Vol. 51 (1929), pp. 363–84.
———. "The Combinatory Foundations of Mathematical Logic," *Journal of Symbolic Logic,* Vol. 7, No. 2 (1942), pp. 49–65.

DANZIG, T. *Number, the Language of Science.* New York, 1930.
DEDEKIND, R. *Essays on Number.* Chicago and London, 1924.
DODD, S. C. *Dimensions of Society.* New York, 1942.
DRESDEN, A. "Brouwer's Contributions to the Foundations of Mathematics," *Bulletin American Mathematical Society,* Vol. 30 (1924), pp. 31–40.
———. "Mathematical Certainty," *Scientia,* Vol. 45 (1929), pp. 369–76.
DUBISLAV, W. "Uber den sogenannten Gegenstand der Mathematik," *Erkenntnis,* Vol. 1 (1930), pp. 27–48.
———. *Die Definition* (3rd edition). Leipzig, 1931.
———. *Die Philosophie der Mathematik in der Gegenwart.* Berlin, 1932.

EATON, RALPH M. *General Logic.* New York, 1931.
ENRIQUES, F. *Historic Development of Logic.* New York, 1929.

FEYS, R. "Les Logiques Nouvelles des modalités," *Revue Néoscolastique de Philosophie,* Vol. 40 (1937), pp. 517–53; Vol. 41 (1938), pp. 217–52.
FINDLAY, J. "Goedelian Sentences: A Non-Numerical Approach," *Mind,* Vol. 51 (1942), pp. 259–66.
FRAENKEL, A. "Axiomatische Theorie der geordneten Mengen," *Journal f. die reine u. angewandte Math.,* Vol. 155 (1926), pp. 129–58.
———. *Einleitung in die Mengenlehre* (3rd edition). Berlin, 1928.
———. "Sur la Notion D'Existence dans les Mathématiques," *L'Enseignement Mathématique,* Vol. 34 (1935), pp. 18–32.
FREGE, GOTTLOB. *Die Grundlagen der Arithmetik.* Breslau, 1884.
———. *Grundgesetze der Arithmetik.* Vol. 1, Jena, 1893. Vol. 2, Jena, 1903.

GAUSS, C. F. *Werke.* Berlin, 1929.
GENTZEN, G. "Neue fassung des Widerspruchsfreiheitsbeweises für die reine Zahlentheorie," in *Forschungen zur Logik,* etc. Heft 4. Leipzig, 1928.
GÖDEL, K. "Uber formal unentscheidbare Sätze der Principia Mathematica und verwandter Systeme," *Monatshefte für Math. und Physik,* Vol. 38 (1931), pp. 173–98.
———. "Uber Vollständigkeit und Widerspruchsfreiheit," *Ergebnisse eines mathematischen Kolloquiums,* Heft 3 (1932), pp. 12–13.
———. *On undecidable propositions of formal mathematical systems.* Princeton, New Jersey, 1934.
GONSETH, F. *Les Mathématiques et la Réalité.* Paris, 1936.
GRAUSTEIN, W. C. *Introduction to Higher Geometry.* New York, 1933.
GREENWOOD, THOMAS. "Invention and Description in Mathematics," *Aristotelian Soc. Proc.* 1929–30 (new series), Vol. 30, pp. 88 ff.

HAHN, HANS. *Logik, Mathematik and Naturerkennen.* Vienna, 1933.
———. *Logique, Mathématique et Connaissance de la Réalité.* Paris, 1935.
HARDY, G. H. "Mathematical Proof," *Mind,* Vol. 38 (1929), pp. 1–25.
———. *Course of Pure Mathematics* (7th edition). Cambridge, 1938.
HAUSDORFF, FELIX. *Mengenlehre.* Berlin and Leipzig, 1927.
HEMPEL, C. G. "A Purely Topological Form of Non-Aristotelian Logic," *Journal of Symbolic Logic,* Vol. 2 (1937), pp. 97–112.
HEYTING, A. "Die formalen Regeln der intuitionistischen Logik," *Sitzungsberichte d. Preussischen Akademie der Wissenschaften,* Phys-Math. Klasse (1930), pp. 42–56.
———. "Die intuitionistische Grundlegung der Mathematik," *Erkenntnis,* Vol. 2 (1931), pp. 106–15.
———. "Mathematische Grundlagenforschungen," *Ergebnisse der Mathematik und ihrer Grenzgebiete,* Vol. 3, No. 4, Berlin, 1934.

HILBERT, David. "Uber den ZahlLegriff," *Jahresbericht der Deutschen Mathematiker-Vereinigung,* Vol. 8, No. 100, pp. 180–84.

———. "The Foundations of Logic and Arithmetic," *Monist,* Vol. 15 (1905), pp. 338–52.

———. "Axiomatisches Denken," *Math. Annalen,* Vol. 78 (1918), pp. 405–15.

———. *Foundations of Geometry* (2nd edition). Chicago, 1921.

———. "Neubegründung der Math.," *Abhandlungan aus dem math. Seminar der Hamburgischen Universitat,* Vol. 1 (1922), pp. 157–77.

———. "Die Logischen Grundlagen der Mathematik," *Math. Annalen,* Vol. 88 (1923), pp. 151–65.

———. "Über das Unendliche," *ibid.,* Vol. 95 (1926), pp. 161–90.

———. "Die Grundlagen der Mathematik," *Abh. Math. Seminar Hamburg Universität,* Vol. 5 (1927), pp. 65–85.

———, and ACKERMAN, W. *Grundzüge d. theoretischen Logik.* Berlin, 1928.

———, and BERNAYS, P. *Grundlagen der Mathematik.* Berlin, 1934.

HOBSON, E. W. *Theory of Functions of a Real Variable,* Vol. 1, (2nd edition). Cambridge, 1921.

HÖLDER, O. *Die Mathematische Methode.* Berlin, 1924.

HUSSERL, E. *Ideas* (trans. by W. R. Boyce Gibson). New York, 1931.

———. *Logische Untersuchungen,* Vol. 1 (4th edition). Halle, 1928.

———. *Formale und Transcendentale Logik.* Halle, 1929.

HUNTINGTON, E. V. "Complete sets of postulates for the theories of positive integral and of positive rational numbers." *Transactions of the American Mathematical Society,* Vol. 3 (1902), pp. 280–84.

———. "Complete sets of postulates for the theory of real quantities," *ibid.,* Vol. 4 (1903), pp. 358–70.

———. *The Continuum.* Cambridge, 1929.

HULL, C. L. *Principles of Behavior: Mathematico-Deductive Theory of Rote Learning.* New York, 1943.

JORDAN, Z. *The Development of Mathematical Logic and of Logical Positivism in Poland between the two Wars.* Polish Science and Learning, London and New York, 1945.

JÖRGENSON, J. *Treatise of Formal Logic* (3 volumes). Copenhagen, 1931.

KATTSOFF, L. O. "Postulational Methods, I," *Philosophy of Science,* Vol. 2 (1935), pp. 139–63.

———. "Postulational Techniques, III," *ibid.,* Vol. 3 (1936), pp. 375–417.

———. "Modality and Probability," *Philosophical Review,* Vol. 46 (1937), pp. 78–85.

———. "Undefined Concepts in Postulate Sets," *ibid.,* Vol. 47 (1938), pp. 293–300.

———. "Philosophy, Psychology, and Postulational Techniques," *Psychological Review,* Vol. 46 (1939), pp. 62–74.

———. "The Relation of Science to Philosophy in the Light of Husserl's Thought," *Philosophical Essays in Memory of E. Husserl.* Harvard University Press, 1941, pp. 203–19.

———. "Physics and Reality," *Philosophy and Phenomenological Research,* Vol. 5 (1944), pp. 108–21.

———. "The Event in Res as Ontological Unit," *Philosophical Review,* Vol. 55 (1946), pp. 174–82.

KAUFMAN, F. *Das Unendliche in der Mathematik, and Seine Ausschaltung.* Leipzig and Vienna, 1930.

KEYSER, C. J. *Mathematical Philosophy.* New York, 1922.

KLEENE, S. C. "On the Interpretation of Intuitionistic Number Theory," *Journal of Symbolic Logic,* Vol. 10 (1945), pp. 109–24.

KLEIN, F. *Elementary Mathematics from an Advanced Standpoint.* New York, 1932.

KORSELT, A. "Was ist Mathematik?" *Archiv. der Mathematik. u. Phys.,* Vol. 21 (1913), pp. 371–73.

KORZYBSKI, A. *Science and Sanity.* Lancaster, Pennsylvania, 1933.

KOYRE, ALEXANDRE. "The Liar," *Philosophy and Phenomenological Research,* Vol. 6 (1946), pp. 344–63.

LALLEMAND. *Le Transfini.* Paris, 1934.
LANGER, S. *Introduction to Symbolic Logic.* New York, 1938.
LANGFORD, C. H. "Analytic Completeness for Postulate Sets," *Proceedings of the London Math. Society,* Vol. 25 (1926), pp. 115–42.
LEFSCHETZ, SOLOMON. "Topology," *American Mathematical Society Colloquium Publication,* Vol. XII, New York, 1930.
LEIBNIZ, G. W. *Philosophische Werke* (Gerhardt edition), Vol. 1.
————. *New Essays Concerning Human Understanding* (trans. A. G. Langley). New York, 1896.
LENZEN, V. F. "Physical Geometry," *American Mathematical Monthly,* Vol. 46 (1939), pp. 324–34.
LEWIS, C. I. *A Survey of Symbolic Logic.* University of California Press, 1918.
————, and LANGFORD, C. H. *Symbolic Logic.* New York, 1932.
LÖWENHEIM, L. "Uber Moglichkeiten im Relative kalkül," *Math. Annalen,* Vol. 76 (1915), pp. 447–70.

MCTAGGERT, J. E. "Propositions Applicable to Themselves," *Mind,* Vol. 32 (1923), pp. 462–64. (Reprinted as Chapter 8 in McTaggert, *Philosophical Studies.* New York, 1934.)
MANIA, BASILIO. "L'infini mathématique et L'évolution de la logique," *Actes du Congrès de Phil. Scientifique.* Paris, Vol. VI, 1936.
MANNOURY, G. "Die signifischen Grundlagen der Mathematik," *Erkenntnis,* Vol. 4 (1934), pp. 288–309, 317–45.
MAUGÉ, F. *L'Esprit et le Réel.* Paris, 1937.
MENGER, K. "Bemerkungen zur Grundlagenfragen," *Jahresb. d. Deutschen Math. Verein.,* Vol. 37 (1928), pp. 213–26, 298–325.
MILL, J. S. *Logic* (8th edition). London, 1919.
MOORE, R. L. "Foundations of Point Set Theory," *American Mathematical Society Colloquium Publications,* Vol. XIII. New York, 1932.
MORRIS, C. W. "Foundations of the Theory of Signs," *International Encyclopedia of Unified Sciences,* Vol. 1, No. 2 and No. 3, Chicago, 1939.
————. *Signs, Language, and Behavior.* New York, 1946.
MÜLLER, A. "Uber Zahlen als Zeichen," *Math. Annalen,* Vol. 90 (1923), pp. 153–58.

NEDER, LUDWIG. "Uber den Aufbau der Arithmetik," *Jahresbericht der Deutschen Mathematiker-Vereinigung,* Vol. 40 (1931), pp. 22 ff.
NEUGEBAUER, O. *Vorgriechische Mathematik.* Berlin, 1934.
NEUMANN, J. VON. "Zur Hilbertschen Beweistheorie," *Math. Zeitschrift,* Vol. 26 (1927), pp. 1–46.
————. "Die formalistische Grundlegung der Mathematik," *Erkenntnis,* Vol. 2 (1931), pp. 116–21.
————, and MORGENSTERN, OSKAR. *Theory of Games and Economic Behavior.* Princeton, 1944.
NICOD, J. "Reduction in Number of Primitive Propositions of Logic," *Proc. Cambridge Phil. Soc.,* Vol. 19 (1917), pp. 32–41.

PAPPERITZ, E. S. "Uber das System der rein mathematischen Wissenschaften," *Jahresb. des Deuts, Math. Verein,* Vol. I. Berlin, 1890–91, pp. 36 ff.
PASCH, M. *Grundlagen der Analysis.* Leipzig, 1909.
————. *Mathematik und Logik.* Leipzig, 1919.
————. *Vorlesungen über neuere Geometrie* (2nd edition). Berlin, 1926.
————. *Der Ursprungdes Zahlbegriffs.* Berlin, 1930.
PEANO, G. *Formulaire de Mathématiques,* Vol. 1. Torino, 1895.
————. "Les definitions mathématiques," *Bibliothèque du Congrès Int. de Phil., Paris, 1900,* Vol. 3, 1903, pp. 279–88.
PEIRCE, C. S. *Collected Papers.* Cambridge, 1933.
POST, E. L. "Introduction to a General Theory of Elementary Propositions," *American Journal of Mathematics,* Vol. 43 (1921), pp. 163–85.

QUINE, W. V. *Mathematical Logic.* New York, 1940.

RAMSEY, F. P. *Foundations of Mathematics.* London, 1931.
RAYNOR, G. E. "Mathematical Induction," *American Mathematical Monthly,* Vol. 33 (1926), pp. 376–77.
REICHENBACH, H. *Wahrscheinlichkeitslehre.* Leyden, 1935.
REYMOND, A. "La négation et le principe du tiers exclu.," *Actes du Congrès intern. de Phil. Scientifique, VI,* Phil. des Math. Paris, 1936, pp. 62 ff.
ROSSER, J. B. "An informal exposition of Proofs of Gödel's Theorems and Church's Theorems," *Journal of Symbolic Logic,* Vol. 4 (1939), pp. 53–60.
RUSSELL, BERTRAND. *Principles of Mathematics* (Vol. 1). Cambridge, 1903 (reprinted in 1937 by W. W. Norton Co.).
————. "Mathematical Logic as Based on the Theory of Types," *American Journal of Mathematics,* Vol. 30 (1908), pp. 222–62.
————. *Introduction to Mathematical Philosophy.* London, 1919.
————, and WHITEHEAD, A. N. *Principia Mathematica:* Vol. 1, 1910, 2nd edition 1925; Vol. 2, 1912, 2nd edition 1927; Vol. 3, 1913, 2nd edition 1927.

SCHOENFLIES, A. "Uber die Stellung der Definition in der Axiomatik," *Jahresb. Deut. Math. Ver.,* Vol. 20 (1911), pp. 222–55.
SCHUBERT, H. C. S. *Mathematical Essays and Recreations.* Chicago, 1898.
SHAW, J. B. *Lectures on Philosophy of Mathematics.* Chicago, 1918.
SMART, H. R. "Frege's Logic," *The Philosophical Review,* Vol. 54 (1945).
SMITH, H. B. "Algebra of Propositions," *Phil. of Science,* Vol. 3 (1936), pp. 551–78.
STAMMLER, G. *Der Zahlbegriff seit Gauss.* Halle, 1926.
STEBBING, L. SUSAN. *Modern Introduction to Logic.* London, 1933.
STOLTZ, D., and GMEINER, J. A. *Theoretische Arithmetik,* Vol. 1. Leipzig and Berlin, 1911.

TARSKI, A. *Einführung in die Math. Logik.* Vienna, 1937. (Also translated as *Introduction to Logic.* New York, 1941.)
TOWNSEND, E. J. *Functions of Real Variables.* New York, 1928.

WAISMANN, F. *Einführung in das Mathematische Denken.* Vienna, 1936.
WEINBERG, J. R. *An Examination of Logical Positivism.* New York, 1936.
WEYL, H. *Philosophie der Mathematik u. Naturwissenschaft.* Munich and Berlin, 1926.
————. "Consistency in Mathematics," *Rice Institute Pamphlets,* No. 16 (1929), pp. 245–65.
————. *Open World.* New Haven, Connecticut, 1931.
————. "The Ghost of Modality," *Philosophical Essays in Memory of E. Husserl.* Cambridge, Massachusetts, (1940), pp. 278–305.
————. "David Hilbert and His Mathematical Work," *Bulletin of American Mathematical Society,* Vol. 50 (1944), pp. 612–55.
WHITEHEAD, A. N. *Universal Algebra.* Cambridge, 1898.
————. *Science and the Modern World.* Macmillan Co., 1928.
————, and RUSSELL, B. *Principia Mathematica* 2nd edition, Vol. 1, Cambridge, 1925; 2nd edition, Vol. 2, Cambridge, 1927; 2nd edition, Vol. 3, Cambridge, 1927.
WOODGER, J. H. *Axiomatic Method in Biology.*

YOUNG, J. W. *Fundamental Concepts of Algebra and Geometry.* New York, 1917.

ZAWIRSKI, Z. "Les Logique Nouvelles et le champ de leur application," *Rev. d. Meta. et de Morale,* Vol. 39 (1932), pp. 503–19.
ZILSEL, EDGAR. "Moritz Schlick," *Naturwissenschaften,* Vol. 25 (1937), pp. 161–67.

SUBJECT INDEX

NAME INDEX